ATHEMATICAL ESEARCH | MATHEMATISCHE FORSCHUNG

Algebraic Methods for Toeplitz-like Matrices and Operators

G. Heinig / K. Rost

Band 19

Akademie-Verlag · Berlin

In this series original contributions of mathematical research in all fields are contained, such as

— research monographs
— collections of papers to a single topic
— reports on congresses of exceptional interest for mathematical research.

This series is aimed at promoting quick information and communication between mathematicians of the various special branches.

In diese Reihe werden Originalbeiträge zu allen Gebieten der mathematischen Forschung aufgenommen wie

— Forschungsmonographien
— Sammlungen von Arbeiten zu einem speziellen Thema
— Berichte von Tagungen, die für die mathematische Forschung besonders aktuell sind.

Die Reihe soll die schnelle Information und gute Kommunikation zwischen den Mathematikern der verschiedenen Fachgebiete fördern.

Manuscripts
in English and German comprising at least 100 pages and not more than 500 pages can be admitted to this series. With respect to a quick publication the manuscripts are reproduced photomechanically. Authors who are interested in this series please turn directly to the 'Akademie-Verlag'. Here you will get more detailed information about the form of the manuscripts and the modalities of publication.

Manuskripte
in englischer und deutscher Sprache, die mindestens 100 Seiten und nicht mehr als 500 Seiten umfassen, können in diese Reihe aufgenommen werden. Im Interesse einer schnellen Publikation werden die Manuskripte auf fotomechanischem Weg reproduziert. Autoren, die an der Veröffentlichung entsprechender Arbeiten in dieser Reihe interessiert sind, wenden sich bitte direkt an den Akademie-Verlag. Sie erhalten dort genauere Informationen über die Gestaltung der Manuskripte und die Modalitäten der Veröffentlichung.

G. Heinig / K. Rost

Algebraic Methods for Toeplitz-like Matrices and Operators

Mathematical Research · Mathematische Forschung

Wissenschaftliche Beiträge
herausgegeben von der
Akademie der Wissenschaften der DDR
Institut für Mathematik

Band 19

Algebraic Methods for Toeplitz-like Matrices and Operators

by
G. Heinig
and
K. Rost

Algebraic Methods for Toeplitz-like Matrices and Operators

by

Georg Heinig
and
Karla Rost

Akademie-Verlag · Berlin 1984

Autoren:
Doz. Dr. sc. nat. Georg Heinig
Dr. rer. nat. Karla Rost
Technische Hochschule Karl-Marx-Stadt
Sektion Mathematik
9010 Karl-Marx-Stadt
Reichenhainer Str. 39/41

Die Titel dieser Schriftenreihe werden vom Originalmanuskript der Autoren reproduziert.

ISSN 0138 - 3019

Erschienen im Akademie-Verlag, DDR-1086 Berlin, Leipziger Str. 3-4

Lizenznummer: 202 • 100/406/84
Printed in the German Democratic Republic
Gesamtherstellung: VEB Kongreß- und Werbedruck, 9273 Oberlungwitz
Lektor: Dr. Reinhard Höppner
Umschlag: Karl Salzbrunn
LSV 1065
Bestellnummer: 763 364 9 (2182/19)
02800

INTRODUCTION

The present volume consists of two parts. The first one is concerned with (finite) matrices having a Toeplitz structure $[a_{i-k}]$ (T-matrices) or a Hankel structure $[b_{i+k}]$ (H-matrices); the second one deals with matrices and linear operators, which are, in some sense, generalizations of T-matrices. Both parts are joined by a common algebraic principle of investigation.

The study of T- and H-matrices has been an active field of research since the beginning of this century and remains it today. The first reason for this circumstance is the fact that such matrices occur in a large variety of areas in pure and applied mathematics. For example, they often appear as discretization of differential and integral equations, they arise in physical data-processing application, in the theories of orthogonal polynomials, stationary processes, moment problems and many others. The second reason is that T- and H-matrices have a lot of significant characteristic properties.

In this monograph we restrict ourselves to algebraic problems concerning T- and H-matrices[1]. We mainly make use of the following three sources:

1) Fast inversion algorithms. As it was shown by N. LEVINSON (in the positive definite case)for the first time, the computational complexity of T-matrix inversion can be reduced from $O(n^3)$ to $O(n^2)$ operations and the storage requirement from $O(n^2)$ to $O(n)$. A lot of papers is dedicated to the further development of this algorithm (see Notes and comments, Part I).

2) Structure of T- and H-matrices. Already E. FROBENIUS proved some theorems about the rank and signature of H-matrices and connections between T- and H-matrices. This direction of investigation has been continued in the papers of I. S. IOHVIDOV and G. HEINIG.

3) Wiener-Hopf theory. The study of finite T-matrices is often motivated by some questions of the Wiener-Hopf theory, which is, in its discrete variant, the theory of infinite T-matrices.

The authors' primary purpose in writing Part I is to present these three aspects under one cover. In this way some new results are obtained published here for the first time. Furthermore, some applications to root localization problems are considered.

Most of the characteristic properties of T-matrices are based on the fact that $AU_n - U_mA$ has at most rank two, if A is an mxn T-matrix and U_n is the forward shift in $\mathbb{C}^n$. This leads to the idea to consider matrices and operators A, for which rank $(AU - VA)$ is small compared

---

[1] For analytic problems see the recent monograph in this series A. BÖTTCHER, B. SILBERMANN [1] or U. GRENANDER, G. SZEGŐ [1].

with rank A, where U and V are two fixed operators. Such operators A will be referred to as <u>UN-Toeplitz-like</u> <u>operators</u> (<u>TLO</u> for short).

In Part II it is shown that many results on finite T-matrices can be generalized to TLO, such as invertibility criteria, inversion formulae and algorithms, kernel structure properties and other. Furthermore, we shall give an abstract definition of the concept of partial indices originated in the Wiener-Hopf theory. We specify the abstract considerations for the following special classes of TLO:

- matrices close to Toeplitz and Hankel (we obtain results similar to those of T. KAILATH et al.)
- matrices close to Vandermonde
- matrices of the form $[(a_i-b_k)^{-1}]$ and their generalizations
- integral operators with kernel close to displacement
- singular integral operators and Toeplitz operators.

Furthermore, we note some applications of our approach to the problem of Wiener-Hopf factorization.

In Part I we hope to draw a quite complete picture of the algebraic theory of T-matrices. On the other hand, the theory of TLO presented in Part II is far from being complete. Our notes offer the current state of the theory and should be viewed as a stepping stone to further development.

We have occasion to thank our colleagues A. Böttcher, U. Jungnickel and B. Silbermann for fruitful conversation. Furthermore, we should like to thank Mrs. M. Graupner for her patient excellent typing work.

Karl-Marx-Stadt, December 1983 The authors

CONTENTS

P A R T I

TOEPLITZ AND HANKEL MATRICES

0. PRELIMINARIES

<u>0.1. Notations.</u> First of all let us introduce and discuss some notations. As usual denote

$\mathbb{N} = \{0,1,\dots\}$ the natural numbers

$\mathbb{Z}$ the integers

$\mathbb{R}$ the real numbers

$\mathbb{C}$ the complex numbers, $\overline{\mathbb{C}} = \mathbb{C} \cup \{\infty\}$

$\mathbb{T} = \{\lambda \in \mathbb{C} : |\lambda| = 1\}$ the unit circle.

If E is a linear space then $E^n = \underbrace{E \times \dots \times E}_{n}$ denotes the cartesian product space, $E^{m\times n}$ denotes the space of mxn matrices with entries from E. The space E^n will be identified with the space of column vectors $E^{n\times 1}$. In this sense $(x_k)_p^q = (x_p \dots x_q)$ always means a column vector. The space of row vectors $E^{1\times n}$ will also be denoted by E_n and $[x_k]_p^q = [x_p \dots x_q]$ always means a row vector. For matrices we use the designation

$$[a_{ik}]_{pr}^{qs} := [a_{ik}]_{\substack{i=p,\dots,q\\ k=r,\dots,s}} ; \; [a_{ik}]_p^q := [a_{ik}]_{pp}^{qq} .$$

The transposed of a matrix A will be denoted by A^T. For sake of simplicity we shall often omit the boundaries of numeration by representing a vector or a matrix in the component form.

For most of our purposes it is convenient to numerate the components of a vector beginning with zero. That means for $x \in \mathbb{C}^n$ we shall numerate $x = (x_k)_0^{n-1}$. Analogously, for mxn matrices we use the numeration $[a_{ik}]_0^{m-1}{}_0^{n-1}$.

By $\{e_0^n,\dots,e_{n-1}^n\}$ we shall denote the canonical basis of $\mathbb{C}^n$

$$e_k^n := (\delta_{ik})_{i=0}^{n-1} , \quad \delta_{ik} := \begin{cases} 0 : i \neq k \\ 1 : i = k \end{cases} .$$

For short we shall often write e_k instead of e_k^n.

In this part we only deal with the complex scalar case $E = \mathbb{C}$. In this case any linear operator $A : \mathbb{C}^n \to \mathbb{C}^m$ corresponds with the matrix in the canonical basis $\mathfrak{A} := [e_i^T A e_k]_0^{m-1}{}_0^{n-1} \in \mathbb{C}^{m\times n}$ and vice versa. In all what follows <u>we shall always identify the operator A</u>

with the corresponding matrix $\mathfrak{X}$.

The kernel and image of an operator A will be denoted by

$$\ker A := \{x : Ax = 0\} \quad \text{and} \quad \operatorname{im} A := \{y = Ax\}.$$

In the following we shall often use the fact that the space $\mathbb{C}^n$ is isomorphic to the space of polynomials with complex coefficients and degree $\leq n-1$. This space will be denoted by $\mathbb{C}^n(\lambda)$. Given $x = (x_k)_0^{n-1} \in \mathbb{C}^n$ denote $x(\lambda) := x_0 + x_1\lambda + \ldots + x_{n-1}\lambda^{n-1} \in \mathbb{C}^n(\lambda)$. Here the variable λ will be regarded as a complex variable. Moreover, we define

$$x(\infty) := x_{n-1}.$$

If $E \subseteq \mathbb{C}^n$ then $E(\lambda)$ will denote the corresponding subset of $\mathbb{C}^n(\lambda)$. Introducing the vectors $l_n(x) := (\lambda^k)_0^{n-1}$ one can write $x(\lambda) = l_n(\lambda)^T x$.

Let us introduce an analogous notation for matrices. Supposing $A = [a_{ik}]_0^{m-1}\,_0^{n-1}$ we define

$$A(\lambda,\mu) := \sum_{i=0}^{m-1} \sum_{k=0}^{n-1} a_{ik}\lambda^i\mu^k .$$

Obviously, $A(\lambda,\mu) = l_m(\lambda)^T A l_n(\mu)$. The polynomial in two variables $A(\lambda,\mu)$ will be called generating function of the matrix A. Some properties of the generating function will be established in Subsection 1.7.

0.2. Toeplitz and Hankel matrices. To any sequence of complex numbers $(a_k)_p^q$ $(p,q \in \mathbb{Z},\ p \leq q)$ we associate the matrices

$$T_{mn}^l(a) := \begin{bmatrix} a_l & a_{l-1} & \cdots & a_{l-n+1} \\ a_{l+1} & a_l & a_{l-1} & \vdots \\ \vdots & a_{l+1} & a_l \ddots & \vdots \\ \vdots & & \ddots & \vdots \\ a_{l+m-1} & & \cdots & a_{l+m-n} \end{bmatrix} \in \mathbb{C}^{m\times n} \tag{0.1}$$

and

$$H_{mn}^l(a) := \begin{bmatrix} a_l & a_{l+1} & a_{l+2} & \cdots & a_{l+n-1} \\ a_{l+1} & a_{l+2} & & & \vdots \\ a_{l+2} & & \cdot^{\cdot^{\cdot}} & & \vdots \\ \vdots & & & & \vdots \\ a_{l+m-1} & & \cdots & & a_{l+m+n-2} \end{bmatrix} \in \mathbb{C}^{m\times n} . \tag{0.2}$$

Here one has to put $a_i := 0$ for $i < p$ and $i > q$. Matrices of the form (0.1) are called Toeplitz matrices ("T-matrices" for short), matrices of the form (0.2) Hankel matrices ("H-matrices" for short). These two classes are the main subject of this part. Furthermore, we shall consider classes of matrices, which are related to the inverses of T- and H-matrices.

For sake of simplicity we set

$$T_{mn}(a) := T^{o}_{mn}(a), \quad T_n(a) := T_{nn}(a),$$

$$H_{mn}(a) := H^{o}_{mn}(a), \quad H_n(a) := H_{nn}(a).$$

The class of $m\times n$ T-matrices will be denoted by $\mathcal{T}(m,n)$, the class of $m\times n$ H-matrices by $\mathcal{H}(m,n)$. In particular, we set $\mathcal{T}(n) := \mathcal{T}(n,n)$, $\mathcal{H}(n) := \mathcal{H}(n,n)$.

Let us introduce some special matrices, which are important for further investigation. Denote

$$I^{l}_{mn} := [\delta_{i,k+l}]_{\substack{i=0,\dots,m-1\\ k=0,\dots,n-1}} \in \mathcal{T}(m,n).$$

We discuss some special cases of the matrices I^{l}_{mn}. Obviously, $I_n := I^{o}_{nn}$ is the $n\times n$ identity matrix;

$$U_n := I^{1}_{nn} = \begin{bmatrix} 0 & & & \\ 1 & \ddots & & \\ & \ddots & \ddots & \\ & & 1 & 0 \end{bmatrix}, \quad U^{T}_n := I^{-1}_{nn} = \begin{bmatrix} 0 & 1 & & \\ & \ddots & \ddots & \\ & & \ddots & 1 \\ & & & 0 \end{bmatrix} \quad {}^{1)}$$

are the matrices of the shift operators in $\mathbb{C}^n$:

$$U_n(x_k)^{n-1}_{o} = (0,x_o,\dots,x_{n-2}), \quad U^{T}_n(x_k)^{n-1}_{o} = (x_1,\dots,x_{n-1},0).$$

Furthermore, in case $m > n$ and $0 \leq l < m-n$ I^{l}_{mn} are imbeddings from $\mathbb{C}^n$ into $\mathbb{C}^m$, in case $m < n$ and $m-n < l \leq 0$ I^{l}_{mn} are restrictions $\mathbb{C}^n$ onto $\mathbb{C}^m$. Finally, let us remark that $e^{n}_{l} = I^{l}_{n1}$.

By J_n we denote the matrix of the counteridentity $J_n(x_k)^{n-1}_{o} = (x_{n-1},\dots,x_o)$:

$$J_n = \begin{bmatrix} & & 1 \\ & 1 & \\ 1 & & \end{bmatrix}.$$

1) Here and in all what follows empty places in a matrix have to be replaced by zeros.

Obviously, $J_n^2 = I_n$. The counteridentity realizes a simple connection between T- and H-matrices.

<u>PROPOSITION 0.1.</u> If $A \in \mathcal{T}(m,n)$ then $AJ_n \in \mathcal{H}(m,n)$ and $J_m A \in \mathcal{H}(m,n)$ and vice versa, $B \in \mathcal{H}(m,n)$ implies $BJ_n \in \mathcal{T}(m,n)$ and $J_m B \in \mathcal{T}(m,n)$.

Some more profound connections between the classes $\mathcal{T}(m,n)$ and $\mathcal{H}(m,n)$ will be studied in Section 4. For the application of the counteridentity we use the designation

$$\hat{x} := J_n x \quad (x \in \mathbb{C}^n).$$

The following proposition shows a relationship between T- or H-matrices and their transposed.

<u>PROPOSITION 0.2.</u> Let $T^l_{mn}(a)$ be defined by (0.1) and $H^l_{mn}(a)$ by (0.2). Then

$$T^l_{mn}(a)^T = J_n T^{m-n+l}_{nm} J_m$$

and

$$H^l_{mn}(a)^T = H^l_{nm}(a).$$

The main tool for all further investigation in this part (one can even say in this book) is the fact that for any T-matrix $A \in \mathcal{T}(m,n)$ the matrix $AU_n - U_m A$ has rank less or equal 2. More precisely speaking, we have

<u>PROPOSITION 0.3.</u> For a T-matrix $A = T^l_{mn}(a)$ the following equalities hold

$$AU_n - U_m A = \begin{bmatrix} a_{l-1} & \cdots & a_{l-n+1} & 0 \\ & & & -a_{l-n+1} \\ & & & \vdots \\ & & & -a_{m-n-1} \end{bmatrix}, \tag{0.3}$$

$$AU_n^T - U_m^T A = \begin{bmatrix} -a_{l+1} & & & \\ \vdots & & & \\ -a_{l+m-1} & & & \\ 0 & a_{l+m-1} & \cdots & a_{m-n+1} \end{bmatrix}. \tag{0.4}$$

The Hankel analogue of Proposition 0.3 is

<u>PROPOSITION 0.3'.</u> For an H-matrix $B = H^l_{mn}(b)$ the following equalities hold

$$BU_n - U_m^T B = \begin{bmatrix} & & & -b_{1+n} \\ & & & \vdots \\ & & & -b_{1+m+n-2} \\ b_{1+m} & \cdots & b_{1+m+n-2} & 0 \end{bmatrix}, \tag{0.5}$$

$$BU_n^T - U_m B = \begin{bmatrix} 0 & b_1 & \cdots & b_{1+n-2} \\ -b_1 & & & \\ \vdots & & & \\ -b_{1+m-2} & & & \end{bmatrix}. \tag{0.6}$$

<u>0.3. Multiplication and difference operators.</u> Let us note some facts about two special classes of rectangular T-matrices, namely the classes of matrices of the form

$$T_{n+r,n}(a) = \begin{bmatrix} a_0 & & \\ \vdots & \ddots & \\ a_r & & a_0 \\ & \ddots & \vdots \\ & & a_r \end{bmatrix} \tag{0.7}$$

and of the transposed form

$$D_n(a) := T_{n+r,n}(a)^T. \tag{0.8}$$

Obviously, the matrices $T_{n+r,n}(a)$ are matrices of the <u>multiplication operator</u> from $\mathbb{C}^n$ into $\mathbb{C}^{n+r}$

$$(T_{n+r,n}(a)x)(\lambda) = a(\lambda)x(\lambda).$$

The operators of the form (0.8) will be called <u>difference operators.</u>

For our further investigation we need a kernel description of difference operators. To do this we use the following fact known from linear algebra: The vector u belongs to $\ker A$ if and only if $y^T u = 0$ for all $y \in \operatorname{im} A^T$.

If A is a difference operator, $A = D_n(a)$, then $\operatorname{im} A^T$ can be easily described in the following way:

Let $\lambda_1, \ldots, \lambda_l$ denote the roots of $a(\lambda) \in \mathbb{C}^{r+1}(\lambda)$ and $\nu_1, \ldots, \nu_l$ their multiplicity. Here, in the case of $a_r = 0$ ∞ is said to be a root of $a(\lambda)$ of multiplicity ν if $a_r = \ldots = a_{r-\nu+1} = 0$, $a_{r-\nu} \neq 0$. Then y belongs to $\operatorname{im} D_n(a)^T$ if and only if $y(\lambda)$ has roots $\lambda_1, \ldots, \lambda_l$ of multiplicity at least $\nu_1, \ldots, \nu_l$.

From these facts a kernel description of difference operators can be obtained. Let us introduce the vectors

$$1_m^k(\lambda) := (k!\binom{i}{k}\lambda^{i-k})_0^m \qquad (\lambda \neq 0, \infty),$$

$$1_m^k(\infty) := e_{m-k}, \; 1_m^k(0) := e_k \qquad (k = 0,1,\dots).$$

with $m := n+r-1$. In the following we often omit the subindex m. Obviously, $y^T 1^k(\lambda_0) = (\frac{d^k}{d\lambda^k} y)(\lambda_0)$. Therefore, we get

PROPOSITION 0.4. The vectors $1^k(\lambda_i)$ $(k = 0,\dots,v_i-1;\; i = 1,\dots,l)$ form a basis of $\ker D_n(a)$.

Let us remark that Prop. 0.4 also can be proved using the relation

$$D_n(a)1^k(\lambda_0) = \sum_{s=0}^{k} \binom{k}{s} (\frac{d^{k-s}a}{d\lambda^{k-s}})(\lambda_0)1^s(\lambda_0), \tag{0.9}$$

which follows from the fact that

$$D_n(a)1^k(\lambda_0) = (\frac{d^k}{d\lambda^k} \lambda^s a(\lambda))(\lambda_0))_{s=0}^{n-1}.$$

1. INVERSION FORMULAE

This section is concerned with the problem how the inverse of a T- or H-matrix can be constructed by the help of the solution of some special (so-called "fundamental") equations. This will be the first step in constructing fast algorithms for T- or H-matrix inversion. The second step, the construction of the fundamental solutions, will be realized in Section 3 for the strongly regular, and in Section 6 for the general case.

It is a quite simple fact known from linear algebra that the inversion of a regular nxn matrix A is equivalent to the solution of n equations $Ax_k = e_k$ $(k = 0,\dots,n-1)$. It is clear that any system of less than n equations cannot determine completely the inverse of A, in general. But if the matrix A is Toeplitz or Hankel it has only $2n-1$ degrees of freedom. Therefore, it can be expected that the inverse is already determined by a system of less than n equations, in the optimal case two equations only. Now we are going to show this fact.

1.1. First inversion variant. The construction of inversion formulae is based upon the relations (0.3) and (0.4). We bring these equalities into another form, more convenient for our purposes. Introduce the (column) vectors

$$g(a,\alpha) := (\alpha, a_{1-n}, \ldots, a_{-1}), \quad g(a,\infty) := e_0 ,$$
$$f(a,\alpha) := (a_1, \ldots, a_{n-1}, \alpha) , \quad f(a,\infty) := e_{n-1}.$$

PROPOSITION 1.1. For $T_n(a) \in \mathcal{T}(n)$ the following relations hold

$$T_n(a)U_n - U_nT_n(a) = e_0\hat{g}(a,\alpha)^T - g(a,\alpha)\,\hat{e}_0^T \quad , \tag{1.1}$$
$$T_n(a)U_n^T - U_n^TT_n(a) = e_{n-1}\hat{f}(a,\alpha)^T - f(a,\alpha)\,\hat{e}_{n-1}^T.^{1)} \tag{1.2}$$

Here α is an arbitrary complex number.

In the following we have to deal with the equations

$$T_n(a)x = e_0, \tag{1.3}$$
$$T_n(a)z(\alpha) = g(a,\alpha). \tag{1.4}$$

These equations will be called fundamental equations, their solutions x and $z(\alpha)$ fundamental solutions.

Instead of equations (1.3), (1.4) one can consider the equations

$$T_n(a)y = e_{n-1}, \tag{1.5}$$
$$T_n(a)w(\alpha) = f(a,\alpha), \tag{1.6}$$

which we shall also call fundamental.

We note that (1.1) and (1.2) can be interpreted as special cases of the equalities

$$T_n(a)U_n-U_nT_n(a) = \sigma(\alpha,\beta)(g(a,\beta)\hat{g}(a,\alpha)^T-g(a,\alpha)\hat{g}(a,\beta)^T), \tag{1.7}$$
$$T_n(a)U_n^T-U_n^TT_n(a) =\sigma(\alpha,\beta)(f(a,\beta)\hat{f}(a,\alpha)^T-f(a,\alpha)\hat{f}(a,\beta)^T), \tag{1.8}$$

where $\alpha,\beta \in \overline{\mathbb{C}}$ arbitrary, $\alpha \neq \beta$ and

$$\sigma(\alpha,\beta) := \begin{cases} (\beta-\alpha)^{-1} & : \alpha,\beta \in \mathbb{C} \\ 1 & : \beta = \infty \\ -1 & : \alpha = \infty \end{cases} . \tag{1.9}$$

Let x, y, $z(\alpha)$ be solutions of the equations introduced above. Then, in view of Prop. 0.2, we have for the transposed matrix $T_n(a)^T$:

---

1) We recall that $\hat{a} := J_na$.

$$T_n(a)^T\hat{x} = e_{n-1}, \quad T_n(a)^T\hat{z}(\alpha) = \hat{g}(a,\alpha),$$

$$T_n(a)\hat{y} = e_o \quad , \quad T_n(a)^T\hat{w}(\alpha) = \hat{f}(a,\alpha).$$

As a simple consequence of Prop. 1.1 we deduce the following properties of the inverse of $T_n(a)$.

PROPOSITION 1.2. Let $T_n(a)$ be a regular T-matrix and $x, z(\alpha)$ and $y, w(\alpha)$ the solutions of the fundamental equations (1.3), (1.4) and (1.5), (1.6), respectively. Then

$$T_n(a)^{-1}U_n - U_nT_n(a)^{-1} = z(\alpha)\hat{x}^T - x\hat{z}(\alpha)^T, \tag{1.10}$$

$$T_n(a)^{-1}U_n^T - U_n^TT_n(a)^{-1} = w(\alpha)\hat{y}^T - y\hat{w}(\alpha)^T. \tag{1.11}$$

Using equalities (1.7), (1.8) we get

PROPOSITION 1.3. For regular $T_n(a) \in \mathcal{T}(n)$ and arbitrary $\alpha, \beta \in \bar{\mathbb{C}}$ with $\alpha \neq \beta$ and (1.9), there is

$$T_n(a)^{-1}U_n - U_nT_n(a)^{-1} = \sigma(\alpha,\beta)(z(\alpha)\hat{z}(\beta)^T - z(\beta)\hat{z}(\alpha)^T), \tag{1.12}$$

$$T_n(a)^{-1}U_n^T - U_n^TT_n(a)^{-1} = \sigma(\alpha,\beta)(w(\alpha)\hat{w}(\beta)^T - w(\beta)\hat{w}(\alpha)^T).$$

Because of Prop. 1.1 and 1.2 it seems to be natural to introduce the following transformations

$$D_+(A) := AU_n - U_nA, \quad D_-(A) := AU_n^T - U_n^TA.$$

We ask how to establish the matrix A, knowing $D_{\pm}(A)$. The answer is based upon the following simple algebraic property.

LEMMA 1.1. Let c, d, u, v be linear operators in suitable spaces such that $cu - vc = d$. Then

$$cu^k - v^kc = \sum_{s=0}^{k-1} v^s du^{k-1-s} \quad (k = 1,2,\dots).$$

The proof is given by induction.

Applying Lemma 1.1 to our case we obtain

$$T_n(a)^{-1}U_n^k = U_n^kT_n(a)^{-1} + \sum_{s=0}^{k-1} U_n^sD_+(T_n(a)^{-1})U_n^{k-1-s}$$

$$(k = 1,2,\dots,n-1).$$

Because of $U_n^ke_o = e_k$ this implies

$$T_n(a)^{-1}e_k = U_n^kx + \sum_{s=0}^{k-1} U_n^sD_+(T_n(a)^{-1})e_{k-1-s}. \tag{1.13}$$

Using equality (1.10) the latter formula can be brought into the form

$$T_n(a)^{-1}e_k = U_n^k x + \sum_{s=0}^{k-1} U_n^s(z(\alpha)\hat{x}^T - x\hat{z}(\alpha)^T)e_{k-1-s} \tag{1.14}$$

and results in the following

THEOREM 1.1. Let $T_n(a) \in \mathcal{T}(n)$ and the equations (1.3), (1.4) be solvable, $x = (x_k)_0^{n-1}$, $z(\alpha) = (z_k)_0^{n-1}$. Then $T_n(a)$ is regular, and its inverse can be represented in the form

$$T_n(a)^{-1} = \begin{bmatrix} z_0 & & & \\ z_1 & z_0 & & \\ \vdots & & \ddots & \\ z_{n-1} & \cdots & & z_0 \end{bmatrix} \begin{bmatrix} x_n & x_{n-1} & \cdots & x_1 \\ & x_n & & \vdots \\ & & \ddots & \vdots \\ & & & x_n \end{bmatrix} - \begin{bmatrix} x_0 & & & \\ x_1 & x_0 & & \\ \vdots & & \ddots & \\ x_{n-1} & \cdots & & x_0 \end{bmatrix} \begin{bmatrix} z_n & z_{n-1} & \cdots & z_1 \\ & z_n & & \vdots \\ & & \ddots & \vdots \\ & & & z_n \end{bmatrix}, \tag{1.15}$$

where $x_n := 0$, $z_n := -1$.

PROOF. Since formula (1.15) is the matrix expression of (1.14), it remains to show that the solvability of the fundamental equations (1.3), (1.4) implies the regularity of $T_n(a)$.

Assume that $T_n(a)$ is singular, there is a nontrivial row vector $\varphi = [\varphi_k]_0^{n-1}$ such that $\varphi T_n(a) = 0$. Using the solvability of (1.3), (1.4) we obtain

$$\begin{aligned} \varphi g(a,\alpha) &= \varphi T_n(a)z(\alpha) = 0, \\ \varphi_0 = \varphi e_0 &= \varphi T_n(a)x = 0. \end{aligned} \tag{1.16}$$

Furthermore, we have in view of (1.1)

$$\varphi U_n T_n(a) = 0.$$

This leads to $\varphi_1 = 0$, according to (1.16). Analogously, we obtain $\varphi U_n^k T_n(a) = 0$ $(k = 0,1,\ldots)$ which implies $\varphi_k = 0$, hence $\varphi \equiv 0$. This contradiction shows the regularity of $T_n(a)$, and the proof is complete. ≡

We discuss now some computational aspects of the formula (1.15). First we note that this formula can be written in the form

$$c_{ik} = \sum_{s=0}^{\min(i,k)} (z_{i-s}x_{n-k+s} - x_{i-s}z_{n-k+s}), \tag{1.17}$$

where $T_n(a)^{-1} = [c_{ik}]_0^{n-1}$. From this explicit expression one deduces the recurrent formula

$$c_{ik} = c_{i-1,k-1} + z_i x_{n-k} - x_i z_{n-k} \quad (i,k = 1,2,\ldots,n-1). \tag{1.18}$$

Supposing we set $c_{ik} := 0$ in case one of the indices i,k does not belong to the index set $\{0,1,\dots,n-1\}$ it is easy to conclude from (1.15) that the latter formula is even true for all $i,k = 0,1,\dots,n$. Provided that $x, z(\alpha)$ are known the calculation of one element c_{ik} by means of (1.18) requires $2\min\{i,k\}$ multiplications. This amount can be still reduced to $2\min\{i,k,n-1-i,n-1-k\}$ if we use for $i+k \geq n$ the reversal recursion from (i,k) to $(i-1,k-1)$. Thus the total number of multiplications for calculating of $T_n(a)^{-1}$ runs to $2(n-1)^2$.

REMARK 1.1. If we use the fundamental solutions of (1.5), (1.6) then an analogous statement of Theorem 1.1 holds, and we have the representation for the inverse of $T_n(a)$ as follows

$$T_n(a)^{-1} = \begin{bmatrix} w_{n-1} & \cdots & w_0 \\ & \ddots & \vdots \\ & & w_{n-1} \end{bmatrix} \begin{bmatrix} 0 & & & \\ y_0 & \ddots & & \\ \vdots & \ddots & \ddots & \\ y_{n-2} & \cdots & y_0 & 0 \end{bmatrix} - \begin{bmatrix} y_{n-1} & \cdots & y_0 \\ & \ddots & \vdots \\ & & y_{n-1} \end{bmatrix} \begin{bmatrix} -1 & & & \\ w_0 & \ddots & & \\ \vdots & \ddots & \ddots & \\ w_{n-2} & \cdots & w_0 & -1 \end{bmatrix}.$$

REMARK 1.2. Formula (1.15) can be written in the following compact form using the notation (0.1)

$$T_n(a)^{-1} = T_n(\underline{z})\, T_n(\hat{\underline{x}})^T - T_n(\underline{x})\, T_n(\hat{\underline{z}})^T,$$

where $\underline{z} = (z_k)_0^n$, $\underline{x} = (x_k)_0^n$, $\hat{\underline{x}} = J_{n+1}\underline{x}$, $\hat{\underline{z}} = J_{n+1}\underline{z}$.

REMARK 1.3. A slight generalization of the formula (1.15) is the following one, which results from (1.12):

$$T_n(a)^{-1} = \sigma(\alpha,\beta)(T_n(\underline{z}(\alpha))T_n(\hat{\underline{z}}(\beta))^T - T_n(\underline{z}(\beta))T_n(\hat{\underline{z}}(\alpha))^T).$$

Now let us deal with H-matrices $H_n(b) \in \mathcal{H}(n)$. Since $J_n H_n(b)$ is Toeplitz all statements and remarks about T-matrices can be easily transferred to H-matrices. Therefore, in principle, we could leave the corresponding formulations to the reader. But there are some methodical reasons such that it seems to be necessary to present some of these formulations here.

PROPOSITION 1.1'. If $H_n(b) \in \mathcal{H}(n)$ then

$$\begin{aligned} H_n(b)U_n - U_n^T H_n(b) &= e_{n-1}h(b,\alpha)^T - h(b,\alpha)e_{n-1}^T, \\ H_n(b)U_n^T - U_n H_n(b) &= e_0 k(b,\alpha)^T - k(b,\alpha)e_0^T, \end{aligned} \tag{1.19}$$

where $h(b,\alpha) := (b_n \dots b_{2n-2}, \alpha)$, $k(b,\alpha) := (\alpha, b_0 \dots b_{n-2})$ with $\alpha \in \mathbb{C}$, arbitrary.

Now we define adequate fundamental equations and fundamental solutions

$$H_n(b)y = e_{n-1}, \quad H_n(b)w(\alpha) = h(b,\alpha), \tag{1.20}$$

$$H_n(b)x = e_0 \quad , \quad H_n(b)z(\alpha) = k(b,\alpha).$$

PROPOSITION 1.2'. Let $H_n(b)$ be a regular H-matrix and $y, w(\alpha)$, $x, z(\alpha)$ the fundamental solutions. Then

$$H_n(b)^{-1}U_n^T - U_nH_n(b)^{-1} = w(\alpha)y^T - yw(\alpha)^T, \tag{1.21}$$

$$H_n(b)^{-1}U_n - U_n^TH_n(b)^{-1} = z(\alpha)x^T - xz(\alpha)^T.$$

Now we formulate a Hankel analogue of Theorem 1.1.

THEOREM 1.1'. For $H_n(b) \in \mathcal{H}(n)$ let the equations (1.20) be solvable. Then $H_n(b)$ is regular and its inverse depends merely on the solutions $y = (y_k)_0^{n-1}$, $w(\alpha) = (w_k)_0^{n-1}$ and can be written in the form

$$H_n(b)^{-1} = \begin{bmatrix} w_0 & & & \\ w_1 & w_0 & & \\ \vdots & & \ddots & \\ w_{n-1} & \cdots & & w_0 \end{bmatrix} \begin{bmatrix} y_1 & y_2 & \cdots & y_n \\ y_2 & & \cdot^{\cdot^{\cdot}} & \\ \vdots & \cdot^{\cdot^{\cdot}} & & \\ y_n & & & \end{bmatrix} - \begin{bmatrix} y_0 & & & \\ y_1 & y_0 & & \\ \vdots & & \ddots & \\ y_{n-1} & \cdots & & y_0 \end{bmatrix} \begin{bmatrix} w_1 & w_2 & \cdots & w_n \\ w_2 & & \cdot^{\cdot^{\cdot}} & \\ \vdots & \cdot^{\cdot^{\cdot}} & & \\ w_n & & & \end{bmatrix}, \tag{1.22}$$

where $y_n := 0$, $w_n := -1$.

Formula (1.22) allows a fast calculation (i.e. calculation with $O(n^2)$ operations) of the elements c_{ik} of the inverse matrix analogously to (1.18)

$$c_{ij} = c_{i-1,j+1} + w_{i-1}y_j - y_{i-1}w_j \quad (i,j = 0,1,\ldots,n)\,. \tag{1.23}$$

1.2. Second inversion variant. The inversion formulae of 1.1 are based on the solution of two equations, where the right side of one equation depends on the matrix. We ask if there are inversion formulae based on the solution of equations with right sides independent of the matrix. In this and the following subsections we present two variants of such inversion formulae. It is remarkable that these two inversion variants do not work for all regular T-matrices. Therefore, we have the alternative: fundamental equations with right sides depending on the matrix, or inversion formulae with an additional condition to the matrix.

In this subsection we try to establish the inverse of a regular T-matrix with the help of the solution of the equations

$$T_n(a)x = e_0, \quad T_n(a)y = e_{n-1}. \tag{1.24}$$

Obviously, x is the first and y is the last column of $T_n(a)^{-1}$. Moreover, by Prop. 0.2 we obtain

$$T_n(a)^T\hat{x} = e_{n-1}, \quad T_n(a)^T\hat{y} = e_0$$

or, which is the same,

$$\hat{x}^T T_n(a) = e_{n-1}^T, \quad \hat{y}^T T_n(a) = e_0^T. \tag{1.25}$$

The latter means $\hat{x}^T$ is the last and $\hat{y}^T$ is the first row of the matrix $T_n(a)^{-1}$. Setting $x = (x_k)_0^{n-1}$, $y = (y_k)_0^{n-1}$ we have in particular

$$x_0 = y_{n-1} \quad ,$$

and the problem of finding $T_n(a)^{-1}$ can be illustrated as a discrete boundary value problem:

$$T_n(a)^{-1} = \begin{bmatrix} x_0=y_{n-1} & y_{n-2} & \cdots & y_1 & y_0 \\ x_1 & & & & y_1 \\ \vdots & & ? & & \vdots \\ x_{n-2} & & & & y_{n-2} \\ x_{n-1} & x_{n-2} & \cdots & x_1 & x_0=y_{n-1} \end{bmatrix} .$$

<u>PROPOSITION 1.4.</u> Suppose the equations (1.24) are solvable and $x_0 \neq 0$ then the fundamental equations (1.3), (1.4) are solvable and

$$z(\alpha) = \frac{1}{x_0}(\beta x - U_n y), \tag{1.26}$$

where $\beta := \hat{g}(a,\alpha)^T y = \alpha x_0 + a_{-1}y_0 + \dots + a_{1-n}y_{n-2}$.

<u>PROOF.</u> According to (1.1) we have

$$D_+(T_n(a))y = e_0\hat{g}(a,\alpha)^T y - g(a,\alpha)\hat{e}_0^T y.$$

Respecting that $U_nT_n(a)y = U_ne_{n-1} = 0$ and $T_n(a)x = e_0$ it follows

$$T_n(a)U_ny = \hat{g}(a,\alpha)^T y \; T_n(a)x - x_0 g(a,\alpha),$$

and we obtain

$$T_n(a)(\beta x - U_n y) = x_0 g(a,\alpha). \tag{1.27}$$

Hence equation (1.4) is solvable and (1.26) holds, provided $x_0 \neq 0$. ≡

REMARK 1.4. The condition $x_0 \neq 0$ is, by Cramer's rule, equivalent to the regularity of the restricted T-matrix $T_{n-1}(a)$.

REMARK 1.5. If $T_n(a)$ is regular and $x_0 = 0$ then, in view of (1.27), the vectors x and $U_n y$ are linearly dependent. It is clear that in such a case $T_n(a)^{-1}$ (and also $T_n(a)$) is not uniquely determined by the solutions x and y.

REMARK 1.6. The example of the matrix

$$\begin{bmatrix} 0 & 0 & 1 \\ 0 & 0 & 0 \\ 1 & 0 & 0 \end{bmatrix}$$

shows that the solvability of (1.24) does not imply the regularity of the given matrix. On the other hand, the example

$$\begin{bmatrix} 0 & 1 \\ 1 & 0 \end{bmatrix}$$

demonstrates that $x_0 \neq 0$ is not necessary for the regularity of the matrix.

As we will see at once, it is useful for the symmetry of inversion formulae to introduce a new notation as

$$x_{k-n+1} := y_k \quad (k = 0,1,\dots,n-1).$$

THEOREM 1.2. Suppose the equations (1.24) are solvable and $x_0 \neq 0$. Then $T_n(a)$ is regular and

$$T_n(a)^{-1} = \frac{1}{x_0}\left\{\begin{bmatrix} x_0 & & \\ \vdots & \ddots & \\ x_{n-1} & \cdots & x_0 \end{bmatrix}\begin{bmatrix} x_0 & \cdots & x_{1-n} \\ & \ddots & \vdots \\ & & x_0 \end{bmatrix} - \begin{bmatrix} x_{-n} & & \\ \vdots & \ddots & \\ x_{-1} & \cdots & x_{-n} \end{bmatrix}\begin{bmatrix} x_n & \cdots & x_1 \\ & \ddots & \vdots \\ & & x_n \end{bmatrix}\right\}, \qquad (1.28)$$

where $x_n := 0$, $x_{-n} := 0$.

PROOF. The regularity of $T_n(a)$ follows immediately from Theorem 1.1 and Prop. 1.4. We use relations (1.10) and (1.26) and carry out a simple computation as

$$D_+(T_n(a)^{-1}) = \frac{1}{x_0}[(\beta x - U_n y)\hat{x}^T - x(\beta\hat{x} - U_n\hat{y})^T]$$
$$= \frac{1}{x_0}[x\hat{y}^T U_n - U_n y\, \hat{x}^T].$$

It remains to apply equation (1.13), and we obtain

$$T_n(a)^{-1}e_k = U_n^k x + \frac{1}{x_0}\sum_{s=0}^{k-1} U_n^s(x\hat{y}^T U_n - U_n y\, \hat{x}^T)e_{k-1-s}$$
$$= \frac{1}{x_0}\left(\sum_{s=0}^{k} x_{-k+s}U_n^s x - x_{n+k-s}U_n^{s+1}y\right),$$

which is equivalent to (1.28). ≡

In case $T_n(a)$ is symmetric we have, in view of Prop. 0.2, $T_n(a)\hat{x} = e_{n-1}$, hence $x_{-k} = x_k$, in case $T_n(a)$ is Hermitian there is $x_{-k} = \overline{x}_k$.

COROLLARY 1.1. Let $T_n(a)$ be symmetric or Hermitian. If the equation $T_n(a)x = e_0$ is solvable and $x_0 \neq 0$ then $T_n(a)$ is regular and $T_n(a)^{-1}$ is determined by (1.28).

The following theorem is a consequence of Theorem 1.2 and includes an inversion formula for the restricted T-matrix $T_{n-1}(a)$.

THEOREM 1.3. Suppose the equations (1.24) are solvable and $x_0 \neq 0$. Then $T_{n-1}(a)$ is regular, and its inverse is determined by

$$T_{n-1}(a)^{-1} = \frac{1}{x_0}\left\{ \begin{bmatrix} x_0 & & \\ \vdots & \ddots & \\ x_{n-2} & \cdots & x_0 \end{bmatrix} \begin{bmatrix} x_0 & \cdots & x_{2-n} \\ & \ddots & \vdots \\ & & x_0 \end{bmatrix} - \begin{bmatrix} x_{1-n} & & \\ \vdots & \ddots & \\ x_{-1} & \cdots & x_{1-n} \end{bmatrix} \begin{bmatrix} x_{n-1} & \cdots & x_1 \\ & \ddots & \vdots \\ & & x_{n-1} \end{bmatrix} \right\}. \quad (1.29)$$

PROOF. Introducing the matrices

$$X_+ := \begin{bmatrix} x_0 & & & \\ x_1 & 1 & & \\ \vdots & & \ddots & \\ x_{n-1} & & & 1 \end{bmatrix}, \qquad X_- := \begin{bmatrix} x_0 & x_{-1} & \cdots & x_{1-n} \\ & 1 & & \\ & & \ddots & \\ & & & 1 \end{bmatrix},$$

we have

$$X_- T_n(a) X_+ = \begin{bmatrix} x_0 & 0 \\ 0 & T_{n-1}(a) \end{bmatrix}.$$

This implies

$$T_n(a)^{-1} = X_+ \begin{bmatrix} \frac{1}{x_0} & 0 \\ 0 & T_{n-1}(a)^{-1} \end{bmatrix} X_- .$$

Let c^l_{ik} denote the entries of the matrix $T_l(a)^{-1}$ $(l=n-1,n)$. Then the latter equality takes out

$$c^n_{ik} = c^{n-1}_{i-1,k-1} + \frac{1}{x_0} x_i x_{-k} \quad (i,k = 0,2,\dots,n-1).^{1)}$$

This results in formula (1.29), and the proof is complete. ≡

Now let us formulate the analogue of Theorem 1.2 for H-matrices.

THEOREM 1.2'. Suppose the solutions $x = (x_k)_0^{n-1}$ and $y = (y_k)_0^{n-1}$ of the equations

$$H_n(b)x = e_0, \quad H_n(b)y = e_{n-1}$$

---

1) We set $c^{n-1}_{-1,k} = c^{n-1}_{k,-1} := 0$ $(k= -1,0,\dots)$

exist and $y_0 \neq 0$. Then $H_n(b)$ is regular, and its inverse can be written in the form

$$H_n(b)^{-1} = \frac{1}{y_0}\left\{\begin{bmatrix} y_0 & & \\ y_1 & y_0 & \\ \vdots & & \ddots \\ y_{n-1} & \cdots & y_0 \end{bmatrix}\begin{bmatrix} x_0 & \cdots & x_{n-1} \\ \vdots & \cdot^{\cdot^{\cdot}} & \\ x_{n-1} & & \end{bmatrix} - \begin{bmatrix} 0 & & & \\ x_0 & 0 & & \\ \vdots & \ddots & \ddots & \\ x_{n-2} & \cdots & x_0 & 0 \end{bmatrix}\begin{bmatrix} y_1 & \cdots & y_n \\ \vdots & \cdot^{\cdot^{\cdot}} & \\ y_n & & \end{bmatrix}\right\}, \quad (1.30)$$

where $y_n := 0$.

1.3. Third inversion variant.

A third inversion variant is based upon the solutions of the equations

$$T_n(a)x = e_0, \quad T_n(a)u = e_1. \quad (1.31)$$

As we shall see later, in a sense the second variant is more specialized in T-matrices and the third variant in H-matrices. Nevertheless, first let us formulate this variant in Toeplitz matrix language.

Before formulating the main theorem of this subsection let us note some simple properties. Obviously, in case $T_n(a)$ is regular x will be the first and u the second column of $T_n(a)^{-1}$. With the same arguments used in 1.2 one can conclude that $\hat{x}^T$ is the last and $\hat{u}^T$ the last but one row of $T_n(a)^{-1}$. Setting $x = (x_k)_0^{n-1}$, $u = (u_k)_0^{n-1}$ we have in particular $u_{n-1} = x_{n-2}$.

The third inversion variant can be illustrated in the following form:

$$T_n(a)^{-1} = \begin{bmatrix} x_0 & u_0 & & \\ \vdots & \vdots & & \\ x_{n-3} & u_{n-3} & & \\ x_{n-2}=u_{n-1} & u_{n-2} & \cdots & u_0 \\ x_{n-1} & x_{n-2}=u_{n-1} & \cdots & x_0 \end{bmatrix}.$$

PROPOSITION 1.5. Suppose the equations (1.31) are solvable and $x_{n-1} \neq 0$. Then the fundamental equations (1.3), (1.4) are solvable and

$$z(\alpha) = \frac{1}{x_{n-1}}(u - U_n x + \gamma x), \quad (1.32)$$

where $\gamma := \alpha x_0 + a_{1-n}x_1 + \dots + a_{-1}x_{n-1}$.

PROOF. According to relation (1.1) we get

$$T_n(a)U_nx = U_nT_n(a)x + \gamma e_o - x_{n-1}g(a,\alpha) = e_1+\gamma T_n(a)x-x_{n-1}g(a,\alpha).$$

Hence ,

$$T_n(a)(\gamma x - U_nx + u) = x_{n-1}\, g(a,\alpha).$$

Therefore, under the assumption $x_{n-1} \neq 0$ the equation (1.4) is solvable and (1.32) holds. ≡

REMARK 1.7. The examples

$$\begin{bmatrix} 0 & 1 & 0 \\ 0 & 0 & 1 \\ 0 & 0 & 0 \end{bmatrix} \quad \text{and} \quad \begin{bmatrix} 0 & 1 & 0 \\ 0 & 0 & 1 \\ 1 & 0 & 0 \end{bmatrix}$$

show that in case $x_{n-1} = 0$ the matrix $T_n(a)$ may be regular or singular. Assuming $T_n(a)$ is regular but $x_{n-1} = 0$ we obtain $u = U_nx-\gamma x$, and it becomes clear that in this case $T_n(a)^{-1}$ is not uniquely determined by the solutions x and u.

THEOREM 1.4. Suppose that for $T_n(a) \in \mathcal{T}(n)$ the equations (1.31) are solvable. Then $T_n(a)$ is regular and

$$T_n(a)^{-1} = \frac{1}{x_{n-1}}\{[x_i x_{n-k-1}]_o^{n-1} + \begin{bmatrix} u_o & & \\ u_1 & u_o & \\ \vdots & & \ddots \\ u_{n-1} & \cdots & u_o \end{bmatrix} \begin{bmatrix} 0 & x_{n-1} & \cdots & x_1 \\ & & \ddots & \vdots \\ & & & x_{n-1} \\ & & & 0 \end{bmatrix}$$

$$- \begin{bmatrix} x_o & & \\ x_1 & x_o & \\ \vdots & & \ddots \\ x_{n-1} & \cdots & x_o \end{bmatrix} \begin{bmatrix} 0 & u_{n-1} & \cdots & u_1 \\ & & \ddots & \vdots \\ & & & u_{n-1} \\ & & & 0 \end{bmatrix}\}. \quad (1.33)$$

PROOF. Because of Prop. 1.5 the regularity of $T_n(a)$ is already proved in Theorem 1.1. A simple computation using the relations (1.10) and (1.32) leads to

$$D_+(T_n(a)^{-1}) = \frac{1}{x_{n-1}}(u\hat{x}^T - U_nx\,\hat{x}^T - x\hat{u}^T + x\hat{x}^TU_n).$$

We put this in equation (1.13), simplify the formula, and obtain (1.33), which completes the proof. ≡

Analogously, it is possible to deduce an inversion formula using the solution y, v of

$$T_n(a)y = e_{n-1}, \quad T_n(a)v = e_{n-2}.$$

Now we are going to consider H-matrices.

<u>THEOREM 1.4'.</u> Suppose that for $H_n(b) \in \mathcal{H}(n)$ the equations

$$H_n(b)y = e_{n-1}, \quad H_n(b)v = e_{n-2}$$

are solvable with $y = (y_k)_0^{n-1}$, $v = (v_k)_0^{n-1}$ and $y_{n-1} \neq 0$. Then $H_n(b)$, $H_{n-1}(b)$ are regular and can be represented in the following form

$$H_n(b)^{-1} = \frac{1}{y_{n-1}}\{[y_i y_k]_0^{n-1} + \begin{bmatrix} v_0 & & & \\ v_1 & v_0 & & \\ \vdots & & \ddots & \\ v_{n-1} & \cdots & & v_0 \end{bmatrix} \begin{bmatrix} y_1 & \cdots & y_{n-1} & 0 \\ \vdots & & \cdot^{\cdot^{\cdot}} & \\ y_{n-1} & \cdot^{\cdot^{\cdot}} & & \\ 0 & & & \end{bmatrix} - \begin{bmatrix} y_0 & & & \\ y_1 & y_0 & & \\ \vdots & & \ddots & \\ y_{n-1} & \cdots & & y_0 \end{bmatrix} \begin{bmatrix} v_1 & \cdots & v_{n-1} & 0 \\ \vdots & & \cdot^{\cdot^{\cdot}} & \\ v_{n-1} & \cdot^{\cdot^{\cdot}} & & \\ 0 & & & \end{bmatrix}\}, \tag{1.34}$$

$$H_{n-1}(b)^{-1} = \frac{1}{y_{n-1}} \left\{ \begin{bmatrix} v_0 & & \\ \vdots & \ddots & \\ v_{n-2} & \cdots & v_0 \end{bmatrix} \begin{bmatrix} y_1 & \cdots & y_{n-1} \\ \vdots & \cdot^{\cdot^{\cdot}} & \\ y_{n-1} & & \end{bmatrix} - \begin{bmatrix} y_0 & & \\ \vdots & \ddots & \\ y_{n-2} & \cdots & y_0 \end{bmatrix} \begin{bmatrix} v_1 & \cdots & v_{n-1} \\ \vdots & \cdot^{\cdot^{\cdot}} & \\ v_{n-1} & & \end{bmatrix} \right\}. \tag{1.35}$$

<u>PROOF.</u> By means of Prop. 0.1 the regularity of $H_n(b)$ and formula (1.34) immediately follow from Theorem 1.4.

Denoting

$$Y := \begin{bmatrix} 1 & & & y_0 \\ & \ddots & & \vdots \\ & & 1 & \vdots \\ & & & y_{n-1} \end{bmatrix}$$

we can proceed analogously to Theorem 1.3. In particular, we have

$$Y^T H_n(b) Y = \begin{bmatrix} H_{n-1}(b) & 0 \\ 0 & y_{n-1} \end{bmatrix},$$

which implies

$$H_n(b)^{-1} = Y \begin{bmatrix} H_{n-1}(b)^{-1} & 0 \\ 0 & \frac{1}{y_{n-1}} \end{bmatrix} Y^T .$$

Designating by c_{ij}^l the entries of the inverse $H_l(b)^{-1}$ $(l=n-1,n)$ the latter equation means

$$c_{ij}^n = c_{ij}^{n-1} + \frac{1}{y_{n-1}} y_i y_j \quad (i,j = 0,1,\ldots,n-1),$$ [1]

---

[1] We set $c_{n-1,j}^{n-1} = c_{i,n-1}^{n-1} = 0$.

and we obtain formula (1.35). This completes the proof. ≡

REMARK 1.8. The condition $y_{n-1} \neq 0$ is, by Cramer's rule, equivalent to the regularity of the restricted matrix $H_{n-1}(b)$.

1.4. Symmetric T-matrices. In this subsection we consider the case of a symmetric T-matrix $T_n(a)$. As Corollary 1.1 shows in this case the inverse is already determined by the solution of only one equation (1.24), provided $T_{n-1}(a)$ is regular. The question is what to do if T_{n-1} is singular. It turns out that also in this case the inverse is determined by the solution of only one equation.

In order to formulate the theorem we need some notations. Supposing $a_n = a_{-n} = \alpha$ we define

$$T_\alpha := T_{n+1}(a), \quad T_\alpha^1 := T_n^1 (a).$$

Then

$$\det T_\alpha = \sigma_2\alpha^2 + \sigma_1\alpha + \sigma_0, \tag{1.36}$$

where

$$\sigma_2 := -\det T_{n-1}(a), \; \sigma_1 := 2(-1)^{n+1}\det T_\alpha^1 \;, \; \sigma_0 := \det T_0.$$

Let α_1, α_2 denote the roots of the quadratic polynomial (1.36). In case $\det T_{n-1}(a) = 0$ we set $\alpha_2 = \infty$.

THEOREM 1.5. If for some $\alpha \neq \alpha_1, \alpha_2$ the equation $T_n(a)z = g(a,\alpha)$ is solvable then $T_n(a)$ is regular, and the inverse can be established by formula (1.15), where $z = (z_k)_0^{n-1}$, $z_n := -1$, and

$$x_k := \frac{1}{\varepsilon}(z_{n-k} + z_0 z_k) \quad (k = 0,\dots,n) \tag{1.37}$$

with

$$\varepsilon := \alpha z_0 + a_{n-1}z_1 + \dots + a_0 z_n.$$

PROOF. Consider the matrix T_α. Obviously, $T_\alpha \underline{z} = \varepsilon e_n$, where $\underline{z} = \begin{bmatrix} z \\ -1 \end{bmatrix}$. From relation (1.36) and $\alpha \neq \alpha_1, \alpha_2$ follows that T_α is regular and, therefore, $\varepsilon \neq 0$. In virtue of Prop. 0.2 and the symmetry of T_α, we have $T_\alpha \hat{\underline{z}} = \varepsilon e_0$. For the vector $\underline{x} = (x_k)_0^n$ defined by (1.37) we obtain $T_\alpha \underline{x} = e_0 + z_0 e_n$. Hence, $x = (x_k)_0^{n-1}$ is a solution of the equation (1.3). It remains to apply Theorem 1.1. ≡

Let us remark a slight generalization of Corollary 1.1.

PROPOSITION 1.6. Suppose the equation

$$T_n(a)w = e_0 + \alpha e_{n-1}$$

is solvable for some $\alpha \neq \pm 1$, $w = (w_k)_0^{n-1}$, and $w_0 \neq \alpha w_{n-1}$. Then $T_n(a)$ is regular and the inverse can be established by (1.28), where

$$x_k = x_{-k} = \frac{1}{1-\alpha^2} (w_k - \alpha w_{n-1-k}).$$

The proof is a straightforward calculation.

Analogously to symmetric T-matrices the inversion of Hermitian T-matrices can be simplified.

1.5. Examples. Now we are going to discuss some examples. The first one is a counterexample to a natural conjecture. Before formulating the conjecture recall that the second and third inversion variants show that under some condition ($x_0 \neq 0$ or $x_{n-1} \neq 0$) the inverse of a T-matrix is uniquely determined by two columns of this matrix (first and last or first and second). Therefore, there is the conjecture that for any regular T-matrix the inverse is determined by a pair of columns.

EXAMPLE 1.1.

$$T_* = \begin{bmatrix} 0 & 0 & 1 & 1 \\ 0 & 0 & 0 & 1 \\ 1 & 0 & 0 & 0 \\ 1 & 1 & 0 & 0 \end{bmatrix}$$

is a regular T-matrix. But it can be easily seen that T_* is not uniquely determined by $T_*^{-1}e_i$, $T_*^{-1}e_k$ for any pair (i,k).

Now let us show how in some cases the fundamental solutions and therefore the inverse can be explicitly evaluated.

EXAMPLE 1.2. Suppose

$$T_n = \begin{bmatrix} 1 & \beta & \cdots & \beta^{n-1} \\ \alpha & \ddots & \ddots & \vdots \\ \vdots & \ddots & \ddots & \beta \\ \alpha^{n-1} & \cdots & \alpha & 1 \end{bmatrix},$$

where $\alpha\beta \neq 1$. It is easy to verify that

$$x = (c, -\alpha c, 0\ldots0), \quad y = (0\ldots0, -\beta c, c),$$

where $c := (1 - \alpha\beta)^{-1}$, are the solutions of

$$T_n x = e_0, \quad T_n y = e_{n-1}.$$

Applying Theorem 1.2 we conclude that T_n is regular and

$$T_n^{-1} = \begin{bmatrix} c & -\beta c & & & \\ -\alpha c & (\alpha\beta+1)c & -\beta c & & \\ & -\alpha c & \ddots & \ddots & \\ & & & (\alpha\beta+1)c & -\beta c \\ & & & -\alpha c & c \end{bmatrix} .$$

Let us remark that in case $\alpha\beta = 1$ T_n has only rank 1.

<u>EXAMPLE 1.3.</u> We consider now Jacobi-Toeplitz matrices

$$T_n = \begin{bmatrix} a_0 & a_{-1} & & \\ a_1 & \ddots & \ddots & \\ & \ddots & \ddots & a_{-1} \\ & & a_1 & a_0 \end{bmatrix} \qquad (a_1 a_{-1} \neq 0).$$

Denote by λ_1, λ_2 the roots of $a_{-1}\lambda + a_0 + a_1\lambda^{-1}$. We shall evaluate the fundamental solutions.

<u>Case $\lambda_1 \neq \lambda_2$.</u> Putting $u_i := (\lambda_i^{k+1})_{k=0}^{n-1}$ $(i=1,2)$ we have

$$T_n u_i = - a_1 e_0 - a_{-1}\lambda_i^{n+1} e_{n-1}.$$

Consequently,

$$T_n(u_1 - u_2) = (\lambda_2^{n+1} - \lambda_1^{n+1}) a_{-1} e_{n-1},$$

$$T_n(\lambda_1^{-n-1} u_1 - \lambda_2^{-n-1} u_2) = (\lambda_2^{-n-1} - \lambda_1^{-n-1}) a_1 e_0.$$

From this we conclude that T_n is regular if and only if $\lambda_1^{n+1} \neq \lambda_2^{n+1}$. In the case of $\lambda_1^{n+1} \neq \lambda_2^{n+1}$ the fundamental solutions x and y yield as

$$y = \frac{1}{(\lambda_2^{n+1} - \lambda_1^{n+1}) a_{-1}} (u_1 - u_2), \quad x = \frac{1}{(\lambda_2^{n+1} - \lambda_1^{n+1}) a_1} (\lambda_1^n u_2 - \lambda_2^n u_1).$$

<u>Case $\lambda_1 = \lambda_2$.</u> Putting $v = ((k+1)\lambda_1^k)_0^{n-1}$ we obtain

$$T_n v = -a_{-1}(n+1)\lambda_1^n e_{n-1}.$$

Hence, T_n is regular, and for the fundamental solutions x and y we have

$$y = - \frac{1}{(n+1)\lambda_1^n a_{-1}} v, \qquad x = \frac{1}{a_1}\left(\frac{\lambda_1}{n+1} v - u_1\right).$$

Since in both cases the condition $x_0 \neq 0$ is fulfilled, the second inversion variant (formula (1.28)) can be applied.

<u>1.6. The transformations Δ and ∇.</u> We showed in the previous sections that the main tool for constructing inversion formulae for

T-matrices are the properties of the transformation

$$D_+ : A \to AU_n - U_nA \quad \text{or} \quad D_- : A \to AU_n^T - U_n^TA.$$

Analogous considerations for H-matrices justify the notations

$$N_0 : A \to AU_n^T - U_nA \quad \text{or} \quad N_1 : A \to AU_n - U_n^TA.$$

We shall develop this idea more in Part II. In this subsection we introduce the transformations Δ and ∇ closed to $D_\pm$ and N_0, N_1, respectively.

DEFINITION 1.1. By Δ, ∇ we denote the transformations

$$\Delta, \nabla : \mathbb{C}^{m\times n} \to \mathbb{C}^{(m+1)\times(n+1)}$$

defined for a matrix $A = [a_{ij}]_0^{m-1}\,{}_0^{n-1}$ by

$$\begin{aligned} \Delta A &= [a_{ik} - a_{i-1,k-1}] \\ \nabla A &= [a_{i-1,k} - a_{i,k-1}]. \end{aligned} \qquad (i = 0,\ldots,m;\ k = 0,\ldots,n)^{1)}$$

Clearly, Δ and ∇ are linear and one-to-one. It is easy to construct a left inverse of Δ and ∇. Indeed, the transformations

$$\begin{aligned} \Gamma B &= [\sum_{s=0}^{\min(i,k)} b_{i-s,k-s}], \\ LB &= [\sum_{s=0}^{\min(n-1-i,k)} b_{i+1+s,k-s}] \end{aligned} \qquad (i = 0,\ldots,m-1;\ k = 0,\ldots,n-1)$$

with $B = [b_{ik}]$ are left inverse to Δ and ∇, respectively. Furthermore, ΔA and ∇A have the structure

$$\Delta A = \begin{bmatrix} * & D_+(A) \\ * & * \end{bmatrix} = \begin{bmatrix} * & * \\ -D_-(A) & * \end{bmatrix},$$

$$\nabla A = \begin{bmatrix} -N_0(A) & * \\ * & * \end{bmatrix} = \begin{bmatrix} * & * \\ * & N_1(A) \end{bmatrix}.$$

The following proposition is easy to verify.

PROPOSITION 1.7. The matrix $B = [b_{ik}]$ belongs to the image of Δ if and only if

$$\sum_{i-k=p} b_{ik} = 0 \quad \text{for } p = 0, \pm 1, \ldots,$$

B belongs to the image of ∇ if and only if

$$\sum_{i+k=p} b_{ik} = 0 \quad \text{for } p = 0, 1, \ldots .$$

---

1) Here we set $a_{i,k} = 0$ for $i \notin \{0,1,\ldots,m-1\}$ or $k \notin \{0,1,\ldots,n-1\}$.

In particular, B belongs to the image of ∇ if B is skew-symmetric, B belongs to the image of Δ if BJ_{n+1} is skew-symmetric. For a T-matrix A the matrix ΔA has rank 4, in general - except, if A is triangular it has rank 2. Moreover, from the inversion formulae of T-matrices the somewhat unexpected fact yields that the rank of ΔA^{-1} is always 2. Analogously, for H-matrices B the rank of ∇B is equal to 4, in general, and the rank of ∇B^{-1} is equal to 2.

REMARK 1.9. It is easily seen that rank $\nabla A \leq 1$ or rank $\Delta A \leq 1$ implies $A = 0$.

Now let $T_n(a) \in \mathcal{T}(n)$ be regular and $x, z(\alpha), y$ be the solutions of (1.3), (1.4), (1.5), respectively. We designate

$$\underline{z} := \begin{bmatrix} z(\alpha) \\ -1 \end{bmatrix}, \quad \underline{x} = \begin{bmatrix} x \\ 0 \end{bmatrix}, \quad \underline{y} = \begin{bmatrix} 0 \\ y \end{bmatrix}.$$

Then, by virtue of (1.10), we have

$$\Delta T_n(a)^{-1} = \underline{z}\,\hat{\underline{x}}^T - \underline{x}\,\hat{\underline{z}}^T. \tag{1.38}$$

In case $T_{n-1}(a)$ is also regular from this follows, according to relation (1.26),

$$\Delta T_n(a)^{-1} = \frac{1}{x_0}(\underline{x}\,\hat{\underline{y}}^T - \underline{y}\,\hat{\underline{x}}^T), \tag{1.39}$$

$$\Delta T_{n-1}(a)^{-1} = \frac{1}{x_0}(x\hat{y}^T - y\hat{x}^T). \tag{1.40}$$

Analogously, for a regular H-matrix $H_n(b) \in \mathcal{H}(n)$ and the corresponding fundamental solutions y, $w(\alpha)$, v we obtain

$$\nabla H_n(b)^{-1} = \underline{y}\,\underline{w}^T - \underline{w}\,\underline{y}^T, \tag{1.41}$$

where $\underline{w} = \begin{bmatrix} w(\alpha) \\ -1 \end{bmatrix}$, $\underline{y} = \begin{bmatrix} y \\ 0 \end{bmatrix}$. Consequently, if $H_{n-1}(b)$ is regular then

$$\nabla H_{n-1}(b)^{-1} = \frac{1}{y_{n-1}}(yv^T - vy^T). \tag{1.42}$$

Furthermore, denoting by y^{n-1} the solution of

$$H_{n-1}(b)y^{n-1} = e_{n-2}$$

the latter equality can be written as

$$\nabla H_{n-1}(b)^{-1} = \frac{1}{y_{n-1}}(y(y^{n-1})^T - y^{n-1}y^T). \tag{1.43}$$

1.7. Representation using the generating function. The formulae above can be written in a neat and compact form using the notation of generating function of a matrix introduced in 0.1. Before doing this let us consider some examples.

For the counteridentity J_n we have

$$J_n(\lambda,\mu) = \sum_{s=0}^{n-1} \lambda^s \mu^{n-1-s} = \frac{\lambda^n - \mu^n}{\lambda - \mu} .$$

For the "shifted counteridentities"

$$J_n^{(l)} = [\delta_{i,l-k}]_0^{n-1} \qquad (l = 0,1,\ldots,2n-2)$$

there is

$$J_n^{(l)}(\lambda,\mu) = \frac{\lambda^{l+1} - \mu^{l+1}}{\lambda - \mu} \qquad \text{if } l \leq n-1$$

and

$$J_n^{(l)}(\lambda,\mu) = \frac{\lambda^n \mu^{l-n+1} - \lambda^{l-n+1}\mu^n}{\lambda - \mu} \qquad \text{if } l \geq n.$$

Therefore, the H-matrix $H_n(b)$, $b = (b_k)_0^{2n-2}$ has the following generating function

$$H_n(b)(\lambda,\mu) = \frac{\lambda b^{(1)}(\lambda) - \mu b^{(1)}(\mu) + \lambda^n b^{(2)}(\mu) - \mu^n b^{(2)}(\lambda)}{\lambda - \mu} ,$$

where $b^{(1)}(\lambda) = \sum_{k=0}^{n-1} b_k \lambda^k$, $\quad b^{(2)}(\lambda) = \sum_{k=n}^{2n-2} b_k \lambda^{k-n+1}$.

For the shift powers U_n^k $(k = 0,1,\ldots,n)$ we obtain

$$U_n^k(\lambda,\mu) = \lambda^k + \lambda^{k+1}\mu + \ldots + \lambda^{n-1}\mu^{n-k-1} = \frac{\lambda^k - \lambda^n \mu^{n-k}}{1 - \lambda\mu} . \tag{1.44}$$

Consequently, for the lower triangular T-matrices $T_n(x)$, $T_{n+1}(x)$ with $x = (x_k)_0^n$ there is

$$T_n(x)(\lambda,\mu) = \frac{x(\lambda) - \lambda^n \hat{x}(\mu)}{1 - \lambda\mu} , \tag{1.45}$$

$$T_{n+1}(x)(\lambda,\mu) = \frac{x(\lambda) - \lambda^{n+1} \mu \hat{x}(\mu)}{1 - \lambda\mu} .$$

Analogously, the generating function for upper triangular T-matrices can be obtained.

We note an important property of the generating function.

PROPOSITION 1.8. Suppose

$$C(\lambda,\mu) = a(\lambda)C'(\lambda,\mu)b(\mu) \in \mathbb{C}^{m,n}(\lambda,\mu),$$

where $C'(\lambda,\mu) \in \mathbb{C}^{m',n'}(\lambda,\mu)$, $a(\lambda) \in \mathbb{C}^{r}(\lambda)$, $b(\lambda) \in \mathbb{C}^{s}(\lambda)$, $r+m' = m+1$, $s+n' = n+1$, then

$$C = D_{m'}(a)^T C' D_{n'}(b) \text{ [1]} \tag{1.46}$$

The proof of Prop. 1.8 is elementary computation.

Let us still note that, obviously,

$$A^T(\lambda,\mu) = A(\mu,\lambda)$$

$$(J_n A J_n)(\lambda,\mu) = A(\lambda^{-1},\mu^{-1})(\lambda\mu)^{n-1}.$$

Next we translate the transformations Δ and ∇ into the language of generating function. Having presented the results of Subsection 1.6 the following propositions are trivial.

PROPOSITION 1.9.

$$(\Delta A)(\lambda,\mu) = (1 - \lambda\mu)\, A(\lambda,\mu),$$

$$(\nabla A)(\lambda,\mu) = (\lambda - \mu)\, A(\lambda,\mu).$$

PROPOSITION 1.10. A matrix B belongs to the image of ∇ if and only if $B(\lambda,\lambda) \equiv 0$, it belongs to the image of Δ if and only if $B(\lambda,\lambda^{-1}) \equiv 0$.

By virtue of Prop. 1.9 the formulae (1.38), (1.39), (1.40) yield the following inversion formulae

$$T_n(a)^{-1}(\lambda,\mu) = \frac{\underline{z}(\lambda)\hat{\underline{x}}(\mu) - \underline{x}(\lambda)\hat{\underline{z}}(\mu)}{1 - \lambda\mu}, \tag{1.47}$$

$$T_n(a)^{-1}(\lambda,\mu) = \frac{1}{x_0}\,\frac{\underline{x}(\lambda)\hat{\underline{y}}(\mu) - \underline{y}(\lambda)\hat{\underline{x}}(\mu)}{1 - \lambda\mu}, \tag{1.47a}$$

$$T_{n-1}(a)^{-1}(\lambda,\mu) = \frac{1}{x_0}\,\frac{x(\lambda)\hat{y}(\mu) - y(\lambda)\hat{x}(\mu)}{1 - \lambda\mu}. \tag{1.47b}$$

For H-matrices we analogously obtain due to (1.41), (1.42), (1.43)

$$H_n(b)^{-1}(\lambda,\mu) = \frac{\underline{y}(\lambda)\underline{w}(\mu) - \underline{w}(\lambda)\underline{y}(\mu)}{\lambda - \mu}, \tag{1.48}$$

$$H_{n-1}(b)^{-1}(\lambda,\mu) = \frac{1}{y_{n-1}}\,\frac{y(\lambda)v(\mu) - v(\lambda)y(\mu)}{\lambda - \mu}, \tag{1.48a}$$

$$H_{n-1}(b)^{-1}(\lambda,\mu) = \frac{1}{y_{n-1}}\,\frac{y(\lambda)y^{n-1}(\mu) - y^{n-1}(\lambda)y(\mu)}{\lambda - \mu}. \tag{1.48b}$$

---

[1] For the definition of $D_n(a)$ see (0.8).

1.8. Determinant representation of the fundamental solutions.

If we determine the fundamental solutions for a given regular T-matrix $T_n(a)$ by means of Cramer's rule we obtain the following representations.

PROPOSITION 1.11. Let $d_n := \det T_n(a) \neq 0$ and x, y the solutions of (1.24), z the solution of (1.4) with $\alpha = a_{-n}$. Then

$$x(\lambda) = \frac{1}{d_n} \det \begin{bmatrix} 1 & \lambda & & \lambda^{n-1} \\ a_1 & a_0 & \cdots & a_{2-n} \\ \vdots & & & \vdots \\ a_{n-1} & a_{n-2} & & a_0 \end{bmatrix},$$

$$y(\lambda) = \frac{1}{d_n} \det \begin{bmatrix} a_0 & a_{-1} & \cdots & a_{1-n} \\ \vdots & & & \vdots \\ a_{n-2} & a_{n-3} & \cdots & a_{-1} \\ 1 & \lambda & & \lambda^{n-1} \end{bmatrix},$$

$$\underline{z}(\lambda) = \frac{(-1)^{n-1}}{d_n} \det \begin{bmatrix} 1 & \lambda & \cdots & \lambda^{n-1} & \lambda^n \\ a_0 & a_{-1} & \cdots & a_{1-n} & a_{-n} \\ \vdots & & & \vdots & \vdots \\ a_{n-1} & \cdots & & a_0 & a_{-1} \end{bmatrix}.$$

2. BEZOUTIANS AND RESULTANT MATRICES

2.1. Characterization of H- and T-matrices.

The Hankel or Toeplitz structure can be characterized by means of the transformations ∇ and Δ.

PROPOSITION 2.1. A matrix $C \in \mathbb{C}^{m \times n}$ is Hankel if and only if

$$I^{-1}_{m-1,m+1} \nabla C \, I^{1}_{n+1,n-1} = 0; \tag{2.1}$$

it is Toeplitz if and only if

$$I^{-1}_{m-1,m+1} \Delta C \, I^{1}_{n+1,n-1} = 0. \tag{2.2}$$

Obviously, (2.1) can be replaced by

$$I^{-1}_{m-1,m} N_0(C) \, I^{1}_{n,n-1} = 0$$

or

$$I^{0}_{m-1,m} N_1(C) \, I^{0}_{n,n-1} = 0.$$

Analogously, (2.2) can be replaced by

$$I^{-1}_{m-1,m}\, D_+(C)\, I^{0}_{n,n-1} = 0$$

or

$$I^{0}_{m-1,m}\, D_-(C)\, I^{1}_{n,n-1} = 0.$$

2.2. Characterization of the inverse of H- and T-matrices.

THEOREM 2.1. A regular matrix A is the inverse of an H-matrix if and only if

$$\operatorname{rank} \nabla A = 2;$$

it is an inverse of a T-matrix if and only if

$$\operatorname{rank} \Delta A = 2.$$

PROOF. The necessity part follows from the results of Subsection 1.6. Suppose now $A \in \mathbb{C}^{n\times n}$, A regular and $\operatorname{rank} \nabla A = 2$. Then there is a representation

$$\nabla A = a_1 b_2^T - b_1 a_2^T$$

for vectors $a_i, b_i \in \mathbb{C}^{n+1}$ $(i=1,2)$. We go over to the generating function

$$(\nabla A)(\lambda,\mu) = a_1(\lambda)b_2(\mu) - b_1(\lambda)a_2(\mu).$$

In view of Prop. 1.9, we have

$$a_1(\lambda)b_2(\lambda) = b_1(\lambda)a_2(\lambda). \tag{2.3}$$

Let $a(\lambda)$ be the greatest common divisor (g.c.d.) of $a_1(\lambda)$ and $a_2(\lambda)$, $b(\lambda)$ the g.c.d. of $b_1(\lambda)$ and $b_2(\lambda)$, $a_i(\lambda) = p_i(\lambda)a(\lambda)$, $b_i(\lambda) = q_i(\lambda)b(\lambda)$ $(i=1,2)$. From (2.3) one concludes $\alpha p_i(\lambda) = q_i(\lambda)$, where $\alpha(\neq 0)$ is a constant. Using Prop. 1.9 we obtain

$$A(\lambda,\mu) = \frac{a_1(\lambda)b_2(\mu) - b_1(\lambda)a_2(\mu)}{\lambda - \mu}$$

$$= \alpha p_1(\lambda)\, \frac{a(\lambda)b(\mu) - b(\lambda)a(\mu)}{\lambda - \mu}\, p_2(\mu).$$

According to Prop. 1.8 this yields the matrix factorization

$$A = D_r(p_1)^T A'\, D_r(p_2)\ ^{1)},$$

where r is the maximal degree of $a(\lambda)$ and $b(\lambda)$, A' is the $r\times r$ matrix with the generating function

$$A'(\lambda,\mu) = \alpha\, \frac{a(\lambda)b(\mu) - b(\lambda)a(\mu)}{\lambda - \mu}.$$

Taking into account the regularity of A it follows that $r = n$, $p_1(\lambda) = p_2(\lambda) = 1$, $a_1(\lambda) = a_2(\lambda) = a(\lambda)$, and $b_1(\lambda) = b_2(\lambda) = \alpha . b(\lambda)$.

---

1) For the definition of $D_r(a)$ see (0.8).

We may assume $\alpha = 1$. Hence

$$\nabla A = (ab^T - ba^T) \quad (a,b \in \mathbb{C}^{n+1}), \tag{2.4}$$

and

$$A(\lambda,\mu) = \frac{a(\lambda)b(\mu) - b(\lambda)a(\mu)}{\lambda - \mu} .$$

In particular, we have $A(\lambda,\mu) = A(\mu,\lambda)$, which means A is symmetric. Supposing $a = (a^i)_0^n$, $b = (b^i)_0^n$ and setting $a' = (a^i)_1^n$, $b' = (b^i)_1^n$ we conclude from (2.4)

$$N_1(A) = AU_n - U_n^T A = a'b'^T - b'a'^T$$

and

$$Ae_0 = z,$$

where $z := b^0 a' - a^0 b'$. We establish that $b^0 \neq 0$ or $a^0 \neq 0$, because $a^0 = b^0 = 0$ implies $Ae_0 = 0$, which contradicts the regularity of A. Assume, for example, $a^0 \neq 0$. Then

$$N_1(A) = \frac{1}{a^0}(za'^T - a'z^T)$$

and, consequently,

$$N_0(A^{-1}) = A^{-1}U_n^T - U_n A^{-1} = he_0^T - e_0 h^T,$$

where $h := \frac{1}{a^0} A^{-1} a'$. Taking into account Prop. 2.1 we obtain that A^{-1} is Hankel.

Thus the Hankel part of the theorem is completely proved. Analogously one can prove the Toeplitz part. $\equiv$

<u>2.3. Bezoutians.</u> Because of Theorem 2.1 it seems to be natural to investigate the class of matrices for which rank $\nabla A \leq 2$ (respectively rank $\Delta A \leq 2$). We denote the class of m×n matrices with this property by $\mathcal{B}_H(m,n)$ (respectively $\mathcal{B}_T(m,n)$). Since in the following we mainly deal with the class $\mathcal{B}_H$ we shall omit the subindex H. Furthermore, we set $\mathcal{B}(n) = \mathcal{B}(n,n)$.

We shall see that the subclass of symmetric matrices of $\mathcal{B}_H$ is a class of matrices appearing in the theory of root separation of algebraic polynomials, namely the class of Bezoutians. In generalization of this fact we call the elements of $\mathcal{B}(m,n)$ <u>H-Bezoutians</u> (or Bezoutians) and the elements of $\mathcal{B}_T(m,n)$ <u>T-Bezoutians.</u>

Suppose $A \in \mathcal{B}(m,n)$ then there are vectors $a_1, b_1 \in \mathbb{C}^{m+1}$ and $a_2, b_2 \in \mathbb{C}^{n+1}$ such that

$$\nabla A = a_1 b_2^T - b_1 a_2^T , \tag{2.5}$$

or, in the language of generating functions,

$$\nabla A(\lambda,\mu) = a_1(\lambda)b_2(\mu) - b_1(\lambda)a_2(\mu).$$

In virtue of Prop. 1.9 this is equivalent to

$$A(\lambda,\mu) = \frac{a_1(\lambda)b_2(\mu) - b_1(\lambda)a_2(\mu)}{\lambda - \mu} \tag{2.6}$$

and, in particular, we have

$$a_1(\lambda)b_2(\lambda) = b_1(\lambda)a_2(\lambda). \tag{2.7}$$

These facts justify the following definition.

DEFINITION 2.1. For given polynomials $a_1(\lambda)$, $b_1(\lambda) \in \mathbb{C}^{m+1}(\lambda)$, $a_2(\lambda)$, $b_2(\lambda) \in \mathbb{C}^{n+1}(\lambda)$ satisfying (2.7) the matrix A with the generating function (2.6) is called the Bezoutian (or H-Bezoutian) of the polynomials $a_1(\lambda)$, $a_2(\lambda)$, $b_1(\lambda)$, $b_2(\lambda)$ and will be denoted by $\mathrm{Bez}\,(a_1,a_2,b_1,b_2)$. In particular, we set

$$\mathrm{Bez}(a,b) = \mathrm{Bez}(a,a,b,b),$$

and shall call Bezoutians of this kind classical. The class of classical Bezoutians will be designated by $\mathcal{B}^0(n)$.

PROPOSITION 2.2. A Bezoutian $A \in \mathcal{B}_H(m,n)$ is classical if and only if $m = n$ and $A = A^T$.

PROOF. The necessity part is obvious. Suppose now A is symmetric and $A \neq 0$. Then ∇A has rank 2 and is skew-symmetric, which means

$$a_1 b_2^T - b_1 a_2^T = a_2 b_1^T - b_2 a_1^T .$$

Hence $a_2 = \alpha_1 a_1 + \beta_1 b_1$, $b_2 = \alpha_2 a_1 + \beta_2 b_1$ for some numbers $\alpha_1,\alpha_2,\beta_1,\beta_2$. This yields after comparing the coefficients $\alpha_2 = \beta_1 = 0$, $\alpha := \alpha_1 = \beta_2$ $(\neq 0)$. Consequently, $a_2 = \alpha a_1$, $b_2 = \alpha b_1$. We obtain

$$\mathrm{Bez}(a_1,a_2,b_1,b_2) = \mathrm{Bez}(\alpha a_1,b_1),$$

and the proposition is proved. ≡

REMARK 2.1. From the proof of Theorem 2.1 one concludes that a regular Bezoutian is always classical and, therefore, symmetric. Moreover, Theorem 2.1 can be formulated in the following form.

COROLLARY 2.1. A regular matrix is the inverse of an H-Bezoutian (T-Bezoutian) if and only if it is an H-matrix (T-matrix).

PROPOSITION 2.3. In order that the Bezoutians $B_1 := \mathrm{Bez}(a_1,a_2,b_1,b_2)$ and $B_2 := \mathrm{Bez}(c_1,c_2,d_1,d_2)$ coincide it is necessary and sufficient that a regular 2x2 matrix Γ exists such that

$$[c_1\ d_1] = [a_1\ b_1]\Gamma, \tag{2.8}$$

$$[c_2\ d_2] = \frac{1}{\det \Gamma}\,[a_2\ b_2]\Gamma. \tag{2.9}$$

PROOF. Suppose $B_1 = B_2$, then $\nabla B_1 = \nabla B_2$. In virtue of (2.5) this is

$$a_1 b_2^T - b_1 a_2^T = c_1 d_2^T - d_1 c_2^T$$

or, which is the same,

$$[a_1 b_1]\begin{bmatrix} 0 & 1 \\ -1 & 0 \end{bmatrix}[a_2 b_2]^T = [c_1 d_1]\begin{bmatrix} 0 & 1 \\ -1 & 0 \end{bmatrix}[c_2 d_2]^T.$$

Hence (2.8) holds for some matrix Γ and

$$[c_2 d_2] = [a_2 b_2]\,\Pi$$

for some matrix Π. Comparing the coefficients we obtain

$$\Gamma \begin{bmatrix} 0 & 1 \\ -1 & 0 \end{bmatrix} \Pi^T = \begin{bmatrix} 0 & 1 \\ -1 & 0 \end{bmatrix},$$

which means

$$\Pi^{-1} = -\begin{bmatrix} 0 & -1 \\ 1 & 0 \end{bmatrix} \Gamma^T \begin{bmatrix} 0 & -1 \\ 1 & 0 \end{bmatrix}. \qquad (2.10)$$

A simple calculation shows that the right side of (2.10) is equal to $(\det \Gamma)\Gamma^{-1}$. This implies (2.9).
On the other side, (2.8) and (2.9) yield (2.10), $\nabla B_1 = \nabla B_2$ and, therefore, $B_1 = B_2$. ≡

COROLLARY 2.2. The Bezoutians Bez(a,b) and Bez(c,d) coincide if and only if a regular 2x2 matrix Γ exists such that $\det \Gamma = 1$ and $[c\ d] = [a\ b]\,\Gamma$.

As an example we evaluate the (classical) Bezoutians of a polynomial

$$a(\lambda) = a_0 + a_1\lambda + \dots + a_n\lambda^n$$

and the powers λ^k $(k=0,1,\dots,n)$. We have

$$\mathrm{Bez}(a,\lambda^k)(\lambda,\mu) = \sum_{i=0}^{n} a_i \frac{\lambda^i\mu^k - \lambda^k\mu^i}{\lambda - \mu}$$

$$= \sum_{i=k+1}^{n} a_i(\lambda^{i-1}\mu^k + \dots + \lambda^k\mu^{i-1}) - \sum_{i=0}^{k-1} a_i(\lambda^{k-1}\mu^i + \dots + \lambda^i\mu^{k-1}),$$

which is equivalent to

$$\mathrm{Bez}(a,\lambda^k) = \left[\begin{array}{ccc:ccc} & & -a_0 & & & \\ & \cdot^{\cdot^{\cdot}} & \vdots & & & \\ -a_0 & \cdots & -a_{k-1} & & & \\ \hdashline & & & a_{k+1} & \cdots & a_n \\ & & & \vdots & \cdot^{\cdot^{\cdot}} & \\ & & & a_n & & \end{array}\right] \begin{array}{l} \left.\vphantom{\begin{array}{c}1\\1\\1\end{array}}\right\} k \\ \left.\vphantom{\begin{array}{c}1\\1\\1\end{array}}\right\} n-k \end{array}.$$

Let us introduce a special notation for the Bezoutians Bez(a,1) and $\mathrm{Bez}(a,\lambda^n)$

$$B(a) := \mathrm{Bez}(a,1), \qquad \tilde{B}(a) := -\mathrm{Bez}(a,\lambda^n).$$

These matrices can be used for the representation of general Bezoutians.
Let us remark that

$$B(a) = H_n^1(a) \quad \text{and} \quad \tilde{B}(a) = -H_n^{1-n}(a).$$

PROPOSITION 2.4. Suppose $a_1(\lambda),\ b_1(\lambda) \in \mathbb{C}^{m+1}(\lambda),\ a_2(\lambda), b_2(\lambda) \in \mathbb{C}^{n+1}(\lambda)$ and $m \geq n$, then

$$\mathrm{Bez}(a_1,a_2,b_1,b_2)=B(a_1)J_m\begin{bmatrix}0\\ \tilde{B}(b_2)\end{bmatrix}-B(b_1)J_m\begin{bmatrix}0\\ \tilde{B}(a_2)\end{bmatrix}\Big\}^{m-n} \tag{2.11a}$$

and

$$\mathrm{Bez}(a_1,a_2,b_1,b_2)=\tilde{B}(b_1)J_m\begin{bmatrix}B(a_2)\\ 0\end{bmatrix}-\tilde{B}(a_1)J_m\begin{bmatrix}B(b_2)\\ 0\end{bmatrix}. \tag{2.11b}$$

PROOF. We have

$$\nabla\, \mathrm{Bez}(a_1,a_2,b_1,b_2) = a_1 b_2^T - b_1 a_2^T.$$

Applying the transformation L (see subsection 1.6) we obtain (2.11a). Similarly, (2.11b) is verified. ≡

Let us collect some obvious properties of the Bezoutians.

PROPOSITION 2.5.

1°. $\mathrm{Bez}(a_2,a_1,b_2,b_1) = \mathrm{Bez}(a_1,a_2,b_1,b_2)^T$

2°. $\mathrm{Bez}(b_1,b_2,a_1,a_2) = -\mathrm{Bez}(a_1,a_2,b_1,b_2)$

3°. For $B(\lambda,\mu) := \mathrm{Bez}(a_1,a_2,b_1,b_2)(\lambda,\mu)$ there is

$$B(\lambda,\lambda) = a_1'(\lambda)b_2(\lambda) - a_2(\lambda)b_1'(\lambda),$$

where the prime denotes the derivative.

The following proposition shows that general Bezoutians can be "reduced" in some sense to classical ones.

PROPOSITION 2.6. Let $a_0(\lambda)$ be the g.c.d. of $a_1(\lambda)$ and $a_2(\lambda)$, $b_0(\lambda)$ the g.c.d. of $b_1(\lambda)$ and $b_2(\lambda)$. Furthermore, let $d(\lambda)$ be an arbitrary common divisor of $a_0(\lambda)$ and $b_0(\lambda)$,

$$a(\lambda) := \frac{a_0(\lambda)}{d(\lambda)}, \quad b(\lambda) := \frac{b_0(\lambda)}{d(\lambda)}.$$

Then

$$\mathrm{Bez}(a_1,a_2,b_1,b_2) = D_r(p)^T \mathrm{Bez}(a,b) D_r(q) \tag{2.12}$$

for some polynomials $p(\lambda)$, $q(\lambda)$.

PROOF. By definition we have

$$a_i(\lambda) = a_0(\lambda)p_i(\lambda), \quad b_i(\lambda) = b_0(\lambda)q_i(\lambda) \quad (i=1,2)$$

for some polynomials p_1, p_2, q_1, q_2, where the pairs p_1, p_2 and q_1, q_2 are disjoint. In virtue of (2.7) we obtain

$$\frac{p_1(\lambda)}{p_2(\lambda)} = \frac{q_1(\lambda)}{q_2(\lambda)},$$

which implies $q_i(\lambda) = \alpha p_i(\lambda)$ $(i=1,2)$ for some $\alpha \in \mathbb{C}$. Consequently, for the generating function $B(\lambda,\mu) := \mathrm{Bez}(a_1,a_2,b_1,b_2)(\lambda,\mu)$ we have

$$B(\lambda,\mu) = \frac{a_1(\lambda)b_2(\mu) - b_1(\lambda)a_2(\mu)}{\lambda - \mu}$$

$$= \alpha p_1(\lambda)d(\lambda)\,\frac{a(\lambda)b(\mu) - b(\lambda)a(\mu)}{\lambda - \mu}\,p_2(\mu)d(\mu).$$

Setting $p(\lambda) := \alpha p_1(\lambda)d(\lambda)$ and $q(\lambda) := p_2(\lambda)d(\lambda)$ we obtain (2.12), according to Prop. 1.8, and the proof is complete. ≡

Our next aim is to show a further connection between Bezoutians and H-matrices, which will be applied in Subsection 2.8.

PROPOSITION 2.7. Suppose

$$w(\lambda) := \frac{b_1(\lambda)}{a_1(\lambda)} = \frac{b_2(\lambda)}{a_2(\lambda)}$$

is a rational function, regular at infinity, and

$$w(\lambda) = w_{-1} + w_0\lambda^{-1} + w_1\lambda^{-2}\ldots$$

is its power series evaluation. Then

$$\mathrm{Bez}(a_1,a_2,b_1,b_2) = B(a_1)H_n(w)[{}^{B(a_2)}_{0}].^{1)} \qquad (2.13)$$

PROOF. We have for the generating function $B(\lambda,\mu)$ of $\mathrm{Bez}(a_1,a_2,b_1,b_2)$

$$B(\lambda,\mu) = a_1(\lambda)\,\frac{w(\mu) - w(\lambda)}{\lambda - \mu}\,a_2(\mu),$$

and

$$\frac{w(\mu) - w(\lambda)}{\lambda - \mu} = \sum_{p,q=0}^{\infty} w_{p+q}\lambda^{-p-1}\mu^{-q-1}.$$

Supposing $a_j(\lambda) = \sum_k a_k^{(j)}\lambda^k$ $(j=1,2)$ this leads to

$$B(\lambda,\mu) = \sum_{i,k}\sum_{p,q=0}^{\infty} a_i^{(1)}w_{p+q}a_k^{(2)}\lambda^{i-p-1}\mu^{k-q-1}$$

$$= \sum_{s,t}\sum_{p,q=0}^{\infty} a_{s+p+1}^{(1)}w_{p+q}a_{q+t+1}^{(2)}\lambda^s\mu^t,$$

which implies (2.13). ≡

2.4. The Bezoutian and the companion matrix. The companion matrix of a monic polynomial

$$a(\lambda) = \sum_{k=0}^{n} a_k\lambda^k \in \mathbb{C}^{n+1}(\lambda) \quad (a_n = 1)$$

is by definition the matrix

$$C(a) := \begin{bmatrix} 0 & 1 & & \\ & & \ddots & \\ & & & 1 \\ -a_0 & -a_1 & \ldots & -a_{n-1} \end{bmatrix}.$$

It is easy to verify that

$$\lambda I_n - C(a) = P(\lambda)\,\mathrm{diag}(a(\lambda),1,\ldots,1)Q(\lambda), \qquad (2.14)$$

where

$$P(\lambda) := \begin{bmatrix} 0 & 1 & & \\ & & \ddots & \\ & & & 1 \\ -1 & -a^1(\lambda) & \ldots & -a^{n-1}(\lambda) \end{bmatrix}, \quad Q(\lambda) := \begin{bmatrix} -1 & & \\ \lambda & \ddots & \\ & \lambda & -1 \end{bmatrix},$$

1) If the degrees of $a_1(\lambda)$ and $a_2(\lambda)$ coincide the zero in the right factor has to be omitted.

$a^k(\lambda) := a_k + a_{k+1}\lambda + \ldots + a_n\lambda^{n-k}$. From (2.14) one concludes that the roots of $a(\lambda)$ coincide with the eigenvalues of $C(a)$.
Furthermore, one observes the relations

$$B(a)C(a) = C(a)^T B(a), \tag{2.15}$$

$$C(a)\tilde{B}(a) = \tilde{B}(a)C(a)^T. \tag{2.16}$$

The companion matrix is connected with the Bezoutian in the following way

$$\mathrm{Bez}(a,\lambda^{k+1}) = \mathrm{Bez}(a,\lambda^k)C(a). \tag{2.17}$$

This relation immediately follows from the equality

$$\mathrm{Bez}(a,\lambda^{k+1})(\lambda,\mu) = B(a,\lambda^k)(\lambda,\mu)\mu - a(\mu)\lambda^k.$$

From (2.17) we obtain the so-called Barnett's formula presented in the next proposition.

PROPOSITION 2.8. Let $b(\lambda)$ be a polynomial of degree $\leq n$, then

$$\mathrm{Bez}(a,b) = B(a)b(C(a)). \tag{2.18}$$

Here by $b(C(a))$ the matrix

$$b(C(a)) = \sum_{k=0}^{n} b_k C(a)^k$$

is denoted.

The spectrum of $b(C(a))$ consists of all numbers $b(\lambda_i)$ $(i=1,2,\ldots,n)$, where λ_i are the roots of $a(\lambda)$. Therefore, formula (2.18) implies the following.

PROPOSITION 2.9. The Bezoutian $\mathrm{Bez}(a,b)$ is a regular matrix if and only if the polynomials $a(\lambda)$ and $b(\lambda)$ have no common roots.

Let us remark that, as is easily seen, Proposition 2.9 also holds for a non-monic polynomial $a(\lambda)$ if one sets $a(\infty) := a_n$ (see 0.3).
The following equality is a slight generalization of Barnett's formula.

PROPOSITION 2.10. Let $a(\lambda)$, $b(\lambda)$, $c(\lambda)$ be polynomials of degree n, m, l, respectively, where $m + l \leq n$. Then

$$\mathrm{Bez}(a,bc) = \mathrm{Bez}(a,c)B(a)^{-1}\mathrm{Bez}(a,b). \tag{2.19}$$

PROOF. Clearly, $(bc)(C(a)) = c(C(a))b(C(a))$. Hence, formula (2.18) implies

$$\mathrm{Bez}(a,bc) = B(a)c(C(a))b(C(a)) = \mathrm{Bez}(a,c)b(C(a)),$$

which yields (2.19). ≡

As a further consequence of Barnett's formula we obtain

PROPOSITION 2.11. Let $a(\lambda)$ and $b(\lambda)$ be monic polynomials of the same degree n. Then

$$\mathrm{Bez}(a,b) = M(C(a)^n - C(b)^n), \tag{2.20}$$

where $M := B(a)J_nB(b) = B(b)J_nB(a)$.

PROOF. According to (2.18), we have

$$\tilde{B}(a) = -\mathrm{Bez}(a,\lambda^n) = -B(a)C(a)^n$$

and

$$\tilde{B}(b) = -\operatorname{Bez}(b,\lambda^n) = -B(b)C(b)^n.$$

Applying (2.11b) and (2.15) we achieve (2.20). ≡

2.5. Resultant matrices. A class of matrices closed to the Bezoutians is the class of resultant matrices. By a resultant matrix is meant a matrix of the form

$$\operatorname{Res}^p(a,b) = \left[\begin{array}{cccccc} a_0 & a_1 & \cdots & a_m & & \\ & \ddots & \ddots & & \ddots & \\ & & a_0 & a_1 & \cdots & a_m \\ b_0 & b_1 & \cdots & b_n & & \\ & \ddots & \ddots & & \ddots & \\ & & b_0 & b_1 & \cdots & b_n \end{array}\right] \begin{array}{l} \}n+p \\ \\ \\ \}m+p \\ \\ \\ \end{array} \qquad (-\min(m,n) < p < \infty).$$

In the square case $p = 0$ we shall omit the index p.

Let us remark that the operator $\lambda_0 I_m - C(a)$ is a special case of the resultant operator

$$\lambda_0 I_m - C(a) = \operatorname{Res}(\lambda_0 - \lambda, \tilde{a}(\lambda_0,\lambda)),$$

where $\tilde{a}(\lambda_0,\lambda) = a_0 + a_1\lambda + \ldots + a_{m-1}\lambda^{m-1} + \lambda_0\lambda^{m-1}$.

A connection between the resultant matrix and the Bezoutian is given, in case $m = n$, by the following two propositions.

PROPOSITION 2.12.

$$\operatorname{Res}(a,b) = P \begin{bmatrix} \operatorname{Bez}(a,b) & 0 \\ 0 & I_n \end{bmatrix} Q,$$

where

$$P = \begin{bmatrix} 0 & I_n \\ B(a)^{-1} & J_n B(b) B(a)^{-1} J_n \end{bmatrix}, \quad Q = \begin{bmatrix} I_n & 0 \\ J_n \tilde{B}(a) & J_n B(a) \end{bmatrix}.$$

The proof is a straightforward calculation.

PROPOSITION 2.13.

$$\operatorname{Res}^p(a_1,b_1)^T \begin{bmatrix} 0 & J_{n+p} \\ -J_{n+p} & 0 \end{bmatrix} \operatorname{Res}^p(a_2,b_2) = \begin{bmatrix} 0 & 0 & -B \\ 0 & 0 & 0 \\ B & 0 & 0 \end{bmatrix} \}p, \qquad (2.21)$$

where $B := \operatorname{Bez}(a_1,a_2,b_1,b_2)$.

PROOF. We compare the generating functions of the two sides of equality (2.14). Let R denote the left side of (2.14). Then

$$\begin{aligned} R(\lambda,\mu) &= l_{2n+p}(\lambda)^T R l_{2n+p}(\mu) \\ &= [a_1(\lambda)\ldots\lambda^{n+p-1}a_1(\lambda), b_1(\lambda)\ldots\lambda^{n+p-1}b_1(\lambda)]\cdot \\ &\quad \cdot [\mu^{n+p-1}b_2(\mu)\ldots b_2(\mu), -\mu^{n+p-1}a_2(\mu)\ldots -a_2(\mu)]^T \\ &= B(\lambda,\mu)(\lambda^{n+p} - \mu^{n+p}). \end{aligned}$$

Clearly, the latter polynomial is just the generating function of the right side of (2.21), and the proof is complete. ≡

2.6. Kernel description of Bezoutians and resultant matrices.

In this subsection we describe the kernel of Bezoutians and resultant matrices in terms of the vectors $l_n^k(\lambda)$ introduced in 0.3.

THEOREM 2.2. Let $a_0(\lambda)$ be the g.c.d. of the polynomials $a_1(\lambda)$ and $a_2(\lambda)$, $b_0(\lambda)$ the g.c.d. of the polynomials $b_1(\lambda)$ and $b_2(\lambda)$, and $d(\lambda)$ the g.c.d. of $a_0(\lambda)$ and $b_0(\lambda)$. Furthermore, let $\lambda_1,\lambda_2,\ldots,\lambda_s$ be the roots of $p(\lambda) := \frac{a_1(\lambda)d(\lambda)}{a_0(\lambda)}$ and $i_1,i_2,\ldots,i_s$ their multiplicities; $\mu_1,\mu_2,\ldots,\mu_t$ denote the roots of $q(\lambda) := \frac{a_2(\lambda)d(\lambda)}{a_0(\lambda)}$ with the multiplicities $j_1,j_2,\ldots,j_t$. Then

$$\ker \operatorname{Bez}(a_1,a_2,b_1,b_2) = \operatorname{lin}\{l_n^k(\lambda_r) : k=0,\ldots,i_r-1;\ r=1,\ldots,s\},$$

$$\ker \operatorname{Bez}(a_1,a_2,b_1,b_2)^T = \operatorname{lin}\{l_m^k(\mu_r) : k=0,\ldots,j_r-1;\ r=1,\ldots,t\}.$$

PROOF. In view of Prop. 2.6 we have

$$B := \operatorname{Bez}(a_1,a_2,b_1,b_2) = \gamma\, D_r(p)^T \operatorname{Bez}(a,b) D_r(q),$$

where γ is some constant, $a(\lambda) := \frac{a_0(\lambda)}{d(\lambda)}$, $b(\lambda) := \frac{b_0(\lambda)}{d(\lambda)}$.

Since $a(\lambda)$ and $b(\lambda)$ have, by definition, no common root the matrix $\operatorname{Bez}(a,b)$ is, according to Prop. 2.9, regular. Therefore, we have

$$\ker B = \ker D_r(q)\ , \quad \ker B^T = \ker D_r(p).$$

It remains to apply Prop. 0.4, and the proof is complete. ≡

COROLLARY 2.3. Let $p(\lambda)$ and $q(\lambda)$ be those from Theorem 2.2. Then the vector y belongs to the image of B if $y(\lambda)$ is a multiple of $p(\lambda)$, it belongs to the image of B^T if $y(\lambda)$ is a multiple of $q(\lambda)$.

COROLLARY 2.4. Using the notation of Theorem 2.2 there are

$$\dim \ker \operatorname{Bez}(a_1,a_2,b_1,b_2) = \sum_{r=1}^{s} i_r,$$

$$\dim \ker \operatorname{Bez}(a_1,a_2,b_1,b_2)^T = \sum_{r=1}^{t} j_r.$$

Now let us determine the kernel of resultant matrices. This could be done using the relationships between resultant matrices and Bezoutians. But we choose a more direct way basing on Prop. 0.3.

THEOREM 2.3. Let $\lambda_1,\ldots,\lambda_t$ be the common roots of the polynomials $a(\lambda) \in \mathbb{C}^{m+1}(\lambda)$ and $b(\lambda) \in \mathbb{C}^{n+1}(\lambda)$ (inclusively ∞, in case the highest order of $a(\lambda)$ and $b(\lambda)$ vanish), $i_1,\ldots,i_t$ their common multiplicity. Then for any p the vectors

$$l^k(\lambda_r) \in \mathbb{C}^{m+n+p} \quad (k = 0,1,\dots,i_r-1;\ r = 1,2,\dots,t)$$

form a basis of $\ker \operatorname{Res}^p(a,b)$

PROOF. Obviously,

$$\ker \operatorname{Res}^p(a,b) = \ker D_{p+n}(a) \cap \ker D_{p+m}(b).$$

By Prop. 0.4 any $u \in \ker D_{p+n}(a)$ admits a representation as follows

$$u = \sum_{r=1}^{q} \sum_{k=o}^{j_r-1} \alpha_{rk}\, l^k(\lambda_r),$$

where α_{rk} are complex numbers, $\lambda_1,\dots,\lambda_q$ roots of $a(\lambda)$, and $j_1,\dots,j_q$ their multiplicities. We may assume

$$\alpha_{r,j_r-1} \neq 0. \tag{2.22}$$

In virtue of (0.9), we conclude

$$D_{p+m}(b)u = \sum_{r=1}^{q} \sum_{k=o}^{j_r-1} \alpha_{rk} \sum_{s=o}^{k} \binom{k}{s}\left(\frac{d^{k-s}b}{d\lambda^{k-s}}\right)(\lambda_r) l^s(\lambda_r)$$

$$= \sum_{r=1}^{q} \sum_{s=o}^{j_r-1} \left(\sum_{k=s}^{j_r-1} \alpha_{rk}\binom{k}{s}\left(\frac{d^{k-s}b}{d\lambda^{k-s}}\right)(\lambda_r)\right) l^s(\lambda_r)\ ^{1)}.$$

Using the fact that the vectors $l^s(\lambda_r)$ are linearly independent we obtain that $u \in \ker \operatorname{Res}^p(a,b)$ if and only if

$$\sum_{k=s}^{j_r-1} \alpha_{rk}\binom{k}{s}\left(\frac{d^{k-s}b}{d\lambda^{k-s}}\right)(\lambda_r) = 0 \quad (s=0,\dots,j_r-1;\ r=1,\dots,q),$$

which implies due to our assumption (2.22)

$$\left(\frac{d^k b}{d\lambda^k}\right)(\lambda_r) = 0 \quad (k=0,\dots,j_r-1;\ r=1,\dots,q).$$

Hence, $q = t$ and $j_r = i_r$ $(r = 1,\dots,t)$ which completes the proof of the theorem. ≡

COROLLARY 2.5. With the notation of Theorem 2.3 we have

$$\dim \ker \operatorname{Res}^p(a,b) = \sum_{r=1}^{t} i_r.$$

Especially, the rank of $\operatorname{Res}^p(a,b)$ is maximal if and only if $a(\lambda)$ and $b(\lambda)$ have no common root.

2.7. Root localization in the upper half-plane.

An important problem in the theory of stability is the problem of counting the number of roots of an algebraic polynomial in a given half-plane.[2] We show in this subsection how the Bezoutian concept can be applied for solving such problems. For convenience we restrict ourselves to the case of

---

1) In case $\lambda_r = \infty$ we have to put $\left(\frac{d^k b}{d\lambda^k}\right)(\infty) := k!\, b_{n-k}$.

2) In case it is the left half-plane this problem is known as Routh-Hurwitz problem.

the upper half-plane. In Section 4 we shall consider the root localization problem for the unit disk - the so-called Schur - Cohn problem. Therefore, we introduce now the concept of inertia of a polynomial not only relative to the real axis but also relative to a general curve.

Let Γ be a curve on the Riemann sphere $\mathbb{C}$ decomposing $\mathbb{C}$ into two domains G_+ and G_-. We denote by $\pi_+(a)$ the number of roots of a given polynomial $a(\lambda)$ in $G_{\pm}$ and by $\pi_0(a)$ the number of roots on Γ - in all cases counting multiplicities. The triple

$$\text{in}_\Gamma\, a(\lambda) := (\pi_+(a), \pi_-(a), \pi_0(a))$$

is called <u>inertia of the polynomial $a(\lambda)$ relative to Γ</u>. In this subsection we deal merely with the case $\Gamma = \mathbb{R}$, G_+ is the upper, G_- is the lower half-plane, and we shall omit the subindex Γ.

Given an Hermitian matrix U the number of positive, negative and zero eigenvalues of U (counting multiplicities) will be denoted by $p_+(U)$, $p_-(U)$, $p_0(U)$, respectively. The triple

$$\text{In } U := (p_+(U), p_-(U), p_0(U))$$

is called <u>inertia of the Hermitian matrix U</u>. Clearly,

$$\text{rank } U = p_+(U) + p_-(U),$$
$$\text{sgn } U = p_+(U) - p_-(U),$$

where $\text{sgn } U$ denotes the signature of U.

The main result of this subsection is the theorem of Hermite.

<u>THEOREM 2.4.</u> Let $a(\lambda)$ be a polynomial and $\overline{a}(\lambda)$ the polynomial with the complex conjugate coefficients. Then for the Hermitian matrix

$$B := \frac{1}{i}\, \text{Bez}(a, \overline{a})$$

the following relations hold

$$\pi_{\pm}(a) = p_{\pm}(B) + \varrho(a), \quad \pi_0(a) = p_0(B) - 2\varrho(a).$$

Here $\varrho(a)$ denotes the number of complex conjugate pairs of roots of $a(\lambda)$ (counting multiplicities).

<u>COROLLARY 2.6.</u> The roots of $a(\lambda)$ completely belong to the upper half-plane if and only if B is positive definite.

For the proof of Theorem 2.4 and some further assertions we make use of Sylvester's well-known inertia theorem. Let us give one possible formulation of this theorem.

<u>LEMMA 2.1.</u> Suppose A is an Hermitian $n\times n$ matrix and

$$A = C^* \tilde{A} C,$$

where C is an $m\times n$ matrix of maximal rank. Then

$$\operatorname{In} A = \operatorname{In} \tilde{A} + (0,0,n-m).$$

REMARK 2.2. Suppose $a(\lambda) = b(\lambda) + ic(\lambda)$, where $b(\lambda)$ and $c(\lambda)$ are polynomials with real coefficients. Then for the matrix B defined in Theorem 2.4 we have

$$B = 2\,\operatorname{Bez}(c,b). \tag{2.23}$$

Without restriction of generality we may suppose now that $a(\lambda)$ is monic. Applying Prop. 2.7 we obtain

$$B = -2\,\operatorname{Bez}(b,c) = B(b)H_n(w)B(b),$$

where $w(\lambda) := -2\,\frac{c(\lambda)}{b(\lambda)} = w_{-1} + w_0\lambda^{-1} + w_1\lambda^{-2} + \dots$.

Taking into account $B(b)^* = B(b)$ and Sylvester's inertia theorem, we conclude from Theorem 2.4.

COROLLARY 2.7. Under the assumptions of Theorem 2.4 there are

$$\pi_\pm(a) = p_\pm(H_n(w)) + \varrho(a),\quad \pi_0(a) = p_0(H_n(w)) - 2\varrho(a).$$

REMARK 2.3. In Section 3 we shall show how to compute the inertia of an H-matrix and of a Bezoutian in the strongly regular case, in Section 6 in the general case.

For proving Theorem 2.4 we need still the following

PROPOSITION 2.14. Suppose $a(\lambda) = a_1(\lambda)a_2(\lambda)$, $b(\lambda) = b_1(\lambda)b_2(\lambda)$ where $a_i(\lambda), b_i(\lambda) \in \mathbb{C}^{n_i+1}(\lambda)$ $(i=1,2)$. Then

$$\operatorname{Bez}(a,b) = \operatorname{Res}(a_2,b_1)^T \begin{bmatrix} \operatorname{Bez}(a_1,b_1) & 0 \\ 0 & \operatorname{Bez}(a_2,b_2) \end{bmatrix} \operatorname{Res}(b_2,a_1) \tag{2.24}$$

PROOF. We have

$$\frac{a(\lambda)b(\mu)-a(\mu)b(\lambda)}{\lambda-\mu} = a_2(\lambda)\,\frac{a_1(\lambda)b_1(\mu)-b_1(\lambda)a_1(\mu)}{\lambda-\mu}\,b_2(\mu)$$
$$+\, b_1(\lambda)\,\frac{a_2(\lambda)b_2(\mu)-b_2(\lambda)a_2(\mu)}{\lambda-\mu}\,a_1(\mu).$$

This implies according to Prop. 1.8

$$\operatorname{Bez}(a,b) = D_{n_1}(a_2)^T \operatorname{Bez}(a_1,b_1)D_{n_1}(b_2)$$
$$+\, D_{n_2}(b_1)^T \operatorname{Bez}(a_2,b_2)D_{n_2}(a_1),$$

which is equivalent to (2.24). $\equiv$

Let us consider the special case of monic $a(\lambda)$ and $b(\lambda)$ of the same degree n and $a_2(\lambda) = \lambda_0 - \lambda$, $b_2(\lambda) = \mu_0 - \lambda$ with $\lambda_0, \mu_0 \in \mathbb{C}$. Then $\operatorname{Bez}(a_2,b_2) = \lambda_0 - \mu_0$ and

$$\operatorname{Res}(b_2,a_1) = \mu_0 I_n - C(\tilde{a}_1(\mu_0)),\quad \operatorname{Res}(a_2,b_1) = \lambda_0 I_n - C(\tilde{b}_1(\lambda_0)),$$

where

$$\tilde{a}_1(\mu_o,\lambda) := a_1(\lambda)+\lambda^{n-1}(\mu_o-\lambda), \quad \tilde{b}_1(\lambda_o,\lambda) := b_1(\lambda)+\lambda^{n-1}(\lambda_o-\lambda).$$

Consequently, (2.24) implies

$$\mathrm{Bez}(a,b)=(\lambda_o I_n-C(\tilde{b}_1(\lambda_o)))^T\begin{bmatrix}\mathrm{Bez}(a_1,b_1) & 0\\ 0 & \lambda_o-\mu_o\end{bmatrix}(\mu_o I_n-C(\tilde{a}_1(\mu_o))). \quad (2.25)$$

The right factor is regular if and only if $a_1(\mu_o) \neq 0$, the left if and only if $b_1(\lambda_o) \neq 0$.

Basing on these facts Prop. 2.9 can be easily verified. Indeed, if $a(\lambda)$ and $b(\lambda)$ have a common root $\lambda_o = \mu_o$, then (2.25) implies that $\mathrm{Bez}(a,b)$ is singular. On the other side, if $a(\lambda)$ and $b(\lambda)$ have no common root, then successive application of (2.25) gives a representation of $\mathrm{Bez}(a,b)$ as a product of regular matrices, which completes the proof. ≡

PROOF OF THEOREM 2.4. Let $d(\lambda)$ be the g.c.d. of the polynomials $a(\lambda)$ and $\bar{a}(\lambda)$, $a(\lambda) = a_o(\lambda)d(\lambda)$, $\bar{a}(\lambda) = \bar{a}_o(\lambda)d(\lambda)$. Clearly, $d(\lambda)$ is a polynomial with real coefficients. Applying Prop. 2.14 we obtain

$$B = \mathrm{Res}(d,\bar{a}_o)^T\begin{bmatrix}-i\mathrm{Bez}(a_o,\bar{a}_o) & 0\\ 0 & 0\end{bmatrix}\mathrm{Res}(d,a_o).$$

According to Corollary 2.5, $\mathrm{Res}(d,a_o)$ is regular. Furthermore, $\mathrm{Res}(d,\bar{a}_o)^T = \mathrm{Res}(d,a_o)^*$ and therefore, Sylvester's inertia theorem can be applied. This yields

$$p_{\pm}(B) = p_{\pm}(B_o), \quad p_o(B) = p_o(B_o) + r,$$

where $B_o := -i\,\mathrm{Bez}(a_o,\bar{a}_o)$ and r is the degree of $d(\lambda)$. By Prop. 2.9 B_o is regular, hence, $p_o(B) = r$. On the other hand, we have $r = \pi_o(a) + 2\varrho(a)$, and the last part of the theorem is proved.

Suppose now λ_1 is a root of $a_o(\lambda)$. Then, by construction, λ_1 is not real. We have for some polynomial $a_1(\lambda)$

$$a_o(\lambda) = a_1(\lambda)(\lambda - \lambda_1), \quad \bar{a}_o(\lambda) = \bar{a}_1(\lambda)(\lambda - \bar{\lambda}_1).$$

The application of Prop. 2.14 to the case $a_2(\lambda) = \lambda - \lambda_1$, $b_1(\lambda)=\bar{a}_1(\lambda)$, $b_2(\lambda) = \lambda - \bar{\lambda}_1$ leads to

$$B_o = R^*\begin{bmatrix}B_1 & 0\\ 0 & i(\bar{\lambda}_1-\lambda_1)\end{bmatrix}R,$$

where $R := \mathrm{Res}(\lambda - \bar{\lambda}_1,a_1)$ and $B_1 := -i\,\mathrm{Bez}(a_1,\bar{a}_1)$. By construction and in view of Corollary 2.5, R is regular. According to Sylvester's theorem, there is

$$\mathrm{In}\,B_o = \begin{cases}\mathrm{In}\,B_1 + (1,0,0) & \text{if } \mathrm{Im}\,\lambda_1 > 0\\ \mathrm{In}\,B_1 + (0,1,0) & \text{if } \mathrm{Im}\,\lambda_1 < 0.\end{cases}$$

The same arguments can be applied if B_o is replaced by B_1 and so on. Finally we obtain a representation

$$B_o = P^* \operatorname{diag}(\operatorname{Im} \lambda_m, \dots, \operatorname{Im} \lambda_1)P, \ ^{1)}$$

where P is a regular matrix and $\lambda_1, \dots, \lambda_m$ are the roots of $a_o(\lambda)$. By Sylvester's theorem this yields the first part of Theorem 2.4. The proof is complete. $\equiv$

Hermite's theorem solves the root localization problem merely in case that the polynomials $a(\lambda)$ and $\bar{a}(\lambda)$, or, which is the same, the polynomials $b(\lambda)$ and $c(\lambda)$, have only trivial g.c.d. Let us discuss now the case that $a(\lambda)$ and $\bar{a}(\lambda)$ have a non-trivial g.c.d. $d(\lambda)$. As it is well-known this polynomial $d(\lambda)$ can be easily determined by the help of Euklid's algorithm. Clearly, $d(\lambda)$ has real coefficients only, and $\pi_+(d) = \varrho(d)$. Therefore, to determine $\varrho(d)$ (and with $\varrho(d)$ the inertia of $a(\lambda)$) we have to solve the root localization problem for real polynomials $d(\lambda)$. Obviously, this problem is solved if we know how to determine the number of real roots of $d(\lambda)$. The next subsection is concerned with this question.

2.8. Root localization for real polynomials. Throughout this subsection $d(\lambda) := d_o + d_1\lambda + \dots + d_r\lambda^r$ will be a monic polynomial with real coefficients. Let $\pi'_o(d)$ denote the number of different real roots of $d(\lambda)$ (i.e. not counting multiplicities) and $\varrho'(d)$ the number of different conjugate complex (non-real) pairs of roots of the polynomial $d(\lambda)$.

THEOREM 2.5. For the Bezoutian $B := \operatorname{Bez}(d, d')$ - the prime designates the derivative - the following relation holds

$$\operatorname{In} B = (\pi'_o(d) - \varrho'(d), \varrho'(d), \delta(d)).$$

Here $\delta(d)$ denotes the degree of the g.c.d. of $d(\lambda)$ and $d'(\lambda)$.

In virtue of Prop. 2.7 and Sylvester's theorem for our purpose the consideration of the Bezoutian can be replaced by the consideration of the H-matrix $H_r(w)$, where

$$w(\lambda) := \frac{d'(\lambda)}{d(\lambda)} = w_o\lambda^{-1} + w_1\lambda^{-2} + \dots .$$

Let $\{\lambda_1, \dots, \lambda_l\}$ denote the set of (different) roots of the polynomial $d(\lambda)$ and $\nu_1, \dots, \nu_l$ their multiplicities. Then

$$w(\lambda) = \frac{\nu_1}{\lambda - \lambda_1} + \dots + \frac{\nu_l}{\lambda - \lambda_l},$$

---

1) As usual, for a complex number λ we denote by $\operatorname{Re} \lambda$ the real and by $\operatorname{Im} \lambda$ the imaginary part of λ.

which implies

$$w(\lambda) = \sum_{i=1}^{l} \sum_{k=0}^{\infty} \nu_i \lambda_i^k \lambda^{-k-1}.$$

The latter means that the coefficients w_k are just the Newton sums

$$w_k = \nu_1\lambda_1^k + \nu_2\lambda_2^k + \dots + \nu_l\lambda_l^k \quad (k=0,1\dots). \tag{2.26}$$

These sums can be easily computed. [1)]
In this way we are led to the following assertion which is equivalent to Theorem 2.5 and is known as Theorem of Jacobi-Borchardt.

<u>THEOREM 2.6.</u> For the H-matrix $H_r(w)$ of Newton sums there is

$$\operatorname{In} H_r(w) = (\pi_0'(d) + \varrho'(d),\ \varrho'(d),\ \delta(d)).$$

<u>PROOF.</u> Given a vector of complex numbers $\lambda = (\lambda_1,\dots,\lambda_l)$ we denote by $V_r(\lambda)$ the Vandermonde matrix

$$V_r(\lambda) := \begin{bmatrix} 1 & 1 & \dots & 1 \\ \lambda_1 & \lambda_2 & \dots & \lambda_l \\ \vdots & \vdots & & \vdots \\ \lambda_1^{r-1} & \lambda_2^{r-1} & \dots & \lambda_l^{r-1} \end{bmatrix}.$$

Then we have

$$H_r(w) = V_r(\lambda)\operatorname{diag}(\nu_1,\dots,\nu_l)\, V_r(\lambda)^T.$$

Making the assumption that the roots $\lambda_1,\dots,\lambda_l$ are ordered in such way that on the first $\pi := \pi_0'(d)$ places stand the real and on the next the complex roots we obtain

$$H_r(w) = V_r(\lambda) D V_r(\lambda)^*, \tag{2.27}$$

where D is the block-diagonal matrix

$$D := \operatorname{diag}(\nu_1,\dots,\nu_\pi, \begin{bmatrix} 0 & \nu_{\pi+2} \\ \nu_{\pi+1} & 0 \end{bmatrix}, \dots, \begin{bmatrix} 0 & \nu_n \\ \nu_{n-1} & 0 \end{bmatrix}).$$

It remains to apply Sylvester's theorem and to take into account that

$$\operatorname{In} \begin{bmatrix} 0 & \alpha \\ \alpha & 0 \end{bmatrix} = (1,1,0) \qquad (\alpha \neq 0).$$

Thus Theorem 2.6 and also Theorem 2.5 are proved. ≡
We note that these theorems solve the root localization problem only in case that the polynomials $d(\lambda)$ and $d'(\lambda)$ are disjoint. If $d(\lambda)$ and $d'(\lambda)$ have a non-trivial g.c.d., say $d_1(\lambda)$, one has to apply Theorem 2.5 to the polynomial $d_1(\lambda)$. Then one has to determine the g.c.d. $d_2(\lambda)$ of $d_1(\lambda)$ and $d_1'(\lambda)$ and to apply the theorem to $d_2(\lambda)$, and so on. After a finite number of steps the root localization problem is solved.

1) One possibility (but not the most effective) is $w_k = \operatorname{trace} C(d)^k$.

2.9. Bezoutian representation via Vandermonde matrices. In this subsection we shall present another useful representation of Bezoutian due to F. I. Lander. Suppose $B := \mathrm{Bez}(a,b) \in \mathcal{B}(n)$ and $\lambda = (\lambda_i)_1^n$, $\mu = (\mu_i)_1^n$, then

$$V_n(\lambda)^T B\, V_n(\mu) = [B(\lambda_i,\mu_k)]_0^{n-1}. \tag{2.28}$$

We are looking for vectors λ and μ such that the latter matrix is diagonal. One can easily verify that a necessary and sufficient condition for this fact is the following one: The numbers λ_i and μ_i coincide and are the roots of a non-trivial linear combination $\alpha a(\lambda) + \beta b(\lambda)$. We obtain the following assertion.

PROPOSITION 2.15. Suppose $a(\lambda),\ b(\lambda) \in \mathbb{C}^{n+1}(\lambda)$ and $c(\lambda) := \alpha a(\lambda) + \beta b(\lambda)$. We assume that $c(\lambda)$ has simple roots $\lambda_1,\dots,\lambda_n$. Then

$$V_n(\lambda)^T \mathrm{Bez}(a,b) V_n(\lambda) = \mathrm{diag}(\gamma_1,\dots,\gamma_n),$$

where $\lambda = (\lambda_i)_1^n$ and

$$\gamma_i = \left(\frac{a(\lambda)}{b(\lambda)}\right)'(\lambda_i)b(\lambda_i)^2 = a'(\lambda_i)b(\lambda_i) - a(\lambda_i)b'(\lambda_i). \tag{2.29}$$

For the proof one has to apply (2.28) and Prop. 2.5.

REMARK 2.4. It can be easily shown that there is a non-trivial linear combination of $a(\lambda)$ and $b(\lambda)$ with exclusively simple roots if and only if $a(\lambda)$ and $b(\lambda)$ have no multiple common roots. In the latter case $a(\lambda) + \beta b(\lambda)$ has only simple roots for all β except a finite number of points.

REMARK 2.5. The Bezoutian can be brought into diagonal form also in the general case if one uses instead of Vandermonde matrices $V_n(\lambda)$ "generalized" Vandermonde matrices, that means matrices with columns of the form $1^k(\lambda_i)$ (see 0.3). But because in this case the corresponding formulae are essentially more complicated we restrict ourselves to the case of simple roots.

From Prop. 2.15 and Cor. 2.1 one concludes

COROLLARY 2.8. A regular H-matrix $H \in \mathcal{H}(n)$ can be represented in the form

$$H = V_n(\lambda)\, D\, V_n(\lambda)^T, \tag{2.30}$$

where D is diagonal.

This statement will be generalized to the case of a general H-matrix in Section 8.

Now we are going to apply Prop. 2.15 to the particular case $b(\lambda) = \overline{a}(\lambda)$, where $a(\lambda)$ is assumed to be monic. For short we shall write $B := -i\,\mathrm{Bez}(a,\overline{a})$. We restrict ourselves to the situation that $a(\lambda)$

has no roots distributed symmetric relative to the real line. (As remarked at the end of 2.7 the general case can be led to this one). Let $b(\lambda)$ be the real, $c(\lambda)$ the imaginary part of $a(\lambda)$, and let $\varepsilon \in \mathbb{R}$ be chosen such that the (real) polynomial $b(\lambda) + \varepsilon c(\lambda)$ has simple roots $\lambda_1,\dots,\lambda_n$ only (we have already mentioned that such ε exists). Suppose $\lambda_1,\dots,\lambda_\pi$ are real and $\lambda_{\pi+1},\lambda_{\pi+2} = \overline{\lambda}_{\pi+1},\dots,$ $\lambda_n = \overline{\lambda}_{n-1}$ are complex. Then, according to (2.29) and (2.23), we have

$$V_n(\lambda)^* B V_n(\lambda) = -2\ \mathrm{diag}(\gamma_1,\dots,\gamma_\pi,\begin{bmatrix} 0 & \gamma_{\pi+2} \\ \gamma_{\pi+1} & 0 \end{bmatrix},\dots,\begin{bmatrix} 0 & \gamma_n \\ \gamma_{n-1} & 0 \end{bmatrix}),$$

where

$$\lambda := (\lambda_i)_1^n,\ \gamma_i := b'(\lambda_i)c(\lambda_i) - b(\lambda_i)c'(\lambda_i) = \left(\frac{b(\lambda)}{c(\lambda)}\right)'(\lambda_i)c(\lambda_i)^2 .$$

From this equality we conclude that the signature of B merely depends on the roots of $b(\lambda) + \varepsilon c(\lambda)$ and, moreover,

$$\mathrm{sgn}\ B = -\sum_{i=1}^{\pi} \mathrm{sgn}\ w'(\lambda_i), \tag{2.31}$$

where $w(\lambda) := \frac{b(\lambda)}{c(\lambda)}$. Assume now $\lambda_1 < \lambda_2 < \dots < \lambda_\pi$ and consider two adjoining roots λ_i, λ_{i+1}. If $w'(\lambda_i)$ and $w'(\lambda_{i+1})$ have the same sign then there must be a pole of odd order of $w(\lambda)$ between λ_i and λ_{i+1} (that means a root of $c(\lambda)$), where $w(\lambda)$ jumps from $-\infty$ to ∞ if $w'(\lambda_i) < 0$ and jumps from ∞ to $-\infty$ otherwise. If $w'(\lambda_i)$ and $w'(\lambda_{i+1})$ have different sign, then between λ_i and λ_{i+1} there are no poles or at most poles of even order. The latter observations admit a compact formulation if one uses the concept of Cauchy index.

<u>DEFINITION 2.2.</u> Let $w(\lambda)$ be a rational function real-valued on the oriented curve Γ. Then the <u>Cauchy index</u> $\underline{I_\Gamma w}$ is, by definition, the difference between the numbers of poles of $w(\lambda)$ on Γ, where $w(\lambda)$ jumps from $-\infty$ to ∞, and the number of poles, where $w(\lambda)$ jumps from ∞ to $-\infty$.

As a consequence of (2.31) we have

<u>PROPOSITION 2.16.</u> Suppose $a(\lambda)$ has no roots symmetric relative to the real axis. Then

$$\mathrm{sgn}\ B = I_{\mathbb{R}} \frac{b(\lambda)}{c(\lambda)} .$$

Comparing this result with Hermite's theorem one obtains

<u>COROLLARY 2.9.</u> The roots of $a(\lambda) := b(\lambda) + ic(\lambda)$ belong to the upper half-plane if and only if its real part $b(\lambda)$ and imaginary part $c(\lambda)$ have only simple real roots and between two roots of $b(\lambda)$ lies one root of $c(\lambda)$.

Furthermore, we conclude from Prop. 2.16 that B is positive definite if and only if $(b(\lambda)c(\lambda)^{-1})' < 0$ $(\lambda \in \mathbb{R})$. This leads to the well-known theorem of Hermite-Biehler.

COROLLARY 2.10. The roots of $a(\lambda)$ belong to the upper half-plane if and only if

$$\frac{1}{i}(\overline{a}(\lambda)a'(\lambda) - a(\lambda)\overline{a}'(\lambda)) > 0$$

for all $\lambda \in \mathbb{R}$.

2.10. Liapunov and Stein equations. In this subsection we shall deal with matrix equations of the form

$$AX - XB = C \tag{2.32}$$

and

$$AXB - X = C. \tag{2.33}$$

Here $A \in \mathbb{C}^{m\times m}$, $B \in \mathbb{C}^{n\times n}$, $C \in \mathbb{C}^{m\times n}$ are given matrices and $X \in \mathbb{C}^{m\times n}$ is asked for. Equations of the form (2.32) are called Liapunov, equations of the form (2.33) Stein equations.

Our first aim is to investigate the connection between the equations (2.32), (2.33) on one side and the Liapunov and Stein equations for corresponding companion matrices on the other side:

$$C(a)^T Y - YC(b) = e_0^m(e_0^n)^T, \tag{2.34}$$

$$C(a)Z - ZC(b)^T = e_{m-1}^m(e_{n-1}^n)^T, \tag{2.35}$$

$$C(a)^T YC(b) - Y = e_0^m(e_0^n)^T, \tag{2.34'}$$

$$C(a)ZC(b)^T - Z = e_{m-1}^m(e_{n-1}^n)^T. \tag{2.35'}$$

Here $a(\lambda)$ and $b(\lambda)$ denote monic polynomials such that $a(A) = 0$ and $b(B) = 0$. For example, one can choose $a(\lambda) = \det(\lambda I_m - A)$ as the characteristic polynomial or $a(\lambda) := m_A(\lambda)$, where $m_A(\lambda)$ denotes the minimal polynomial of A.

After then we shall see that the equations (2.35), (2.35') have some connection to the Bezoutians and resultant matrices.

Let us introduce the notation of tensor or Kronecker product of two matrices $A \in \mathbb{C}^{m\times m}$ and $B \in \mathbb{C}^{n\times n}$. Identifying matrices and operators we mean by $A \otimes B$ the linear operator acting in $\mathbb{C}^{m\times n}$ according to the formula

$$(A \otimes B)X := AXB.$$

To any polynomial in two variables

$$p(\lambda,\mu) := \sum p_{ik}\, \lambda^i \mu^k$$

we associate the operator

$$p(A,B) := \sum p_{ik}A^i \otimes B^k.$$

Using this notation the equations (2.32), (2.33) can be written in the form

$$p(A,B)X = C,$$

where $p(\lambda,\mu) := \lambda - \mu$ in the Liapunov case and $p(\lambda,\mu) := \lambda\mu - 1$ in the Stein case. The idea of solving (2.32) or (2.33) due to R. A. Smith and P. C. Müller [1] is to look for the inverse of $A \otimes I - I \otimes B$ or $A \otimes B - I \otimes I$ as operators of the form $y(A,B)$, where $y(\lambda,\mu)$ is a polynomial.

PROPOSITION 2.17. Let $Y = [y_{ik}] \in \mathbb{C}^{m\times n}$ be a solution of (2.34). Then (2.32) is uniquely solvable for all $C \in \mathbb{C}^{m\times n}$, and its solution is determined by

$$X = \sum_{i=0}^{m-1} \sum_{k=0}^{n-1} y_{ik}A^iCB^k. \tag{2.36}$$

PROOF. We have

$$\begin{aligned} AX - XB &= \sum_{i=0}^{m-1} \sum_{k=0}^{n-1} y_{ik}(A^{i+1}CB^k - A^iCB^{k+1}) \\ &= \sum_{i=0}^{m-1} \sum_{k=0}^{n-1} (y_{i-1,k} - y_{i,k-1} - a_i y_{m-1,k} + y_{i,n-1}b_k)A^iCB^k, \end{aligned}$$

where $a(\lambda) := \sum a_i\lambda^i$, $b(\lambda) := \sum b_k\lambda^k$ the polynomials introduced at the beginning of this subsection. Taking into account that $[y_{ik}]$ is a solution of (2.34) we obtain $AX - XB = C$. The solvability of (2.32) for all C implies the uniqueness of the solution, and the proof is complete. ≡

Analogously, the following proposition holds.

PROPOSITION 2.17'. Let $Y = [y_{ik}] \in \mathbb{C}^{m\times n}$ be a solution of (2.34'). Then (2.33) is uniquely solvable for all $C \in \mathbb{C}^{m\times n}$, and its solution is determined by

$$X = \sum_{i=0}^{m-1} \sum_{k=0}^{n-1} y_{ik}A^iCB^k.$$

REMARK 2.6. Prop. 2.17 clearly includes the following assertion. If (2.34) is solvable, then $A \otimes I_n - I_m \otimes B$ is invertible. If we take, in particular, $A = C(a)^T$, $B = C(b)$ we obtain the statement that in case $e_0^m(e_0^n)^T$ belongs to the image of $C(a)^T \otimes I_n - I_m \otimes C(b)$ this operator is invertible.

Let us now investigate the connection between (2.34) and (2.35) (respectively (2.34') and (2.35')). First we recall the relation (2.15)

---

[1] See also "Notes and comments".

$$B(a)C(a) = C(a)^T B(a).$$

From this equality we obtain that

$$Y = B(a)ZB(b)$$

is a solution of (2.34) if Z is a solution of (2.35) and, vice versa, if Y is a solution of (2.34) then $Z = B(a)^{-1}YB(b)^{-1}$ is a solution of (2.35).

Since $a(\lambda)$ is a multiple of the minimal polynomial of A, $a(\lambda)(\lambda I_m - A)^{-1}$ will be a matrix polynomial. We denote by A_i the coefficient of λ^i of this matrix polynomial and analogously by B_i the coefficients of $b(\lambda)(\lambda I_n - B)^{-1}$. From the series expansion

$$\sum_i A_i\lambda^i = a(\lambda)(\lambda^{-1}I_m + \lambda^{-2}A + \dots)$$

immediately follows the relation

$$A_i = \sum_{k=0}^{m-i-1} a_{i+k+1}A^k \quad (i=0,1,\dots,m-1)$$

(analogously for B_i).

We remark that a_{i+k+1} are just the elements of the matrix $B(a)$. Therefore, if the matrices $Y = [y_{ik}]$ and $Z = [z_{ik}]$ are connected via $Y = B(a)ZB(b)$, then

$$\sum_{i,k} y_{ik}A^i \otimes B^k = \sum_{i,k}\sum_{p,q} a_{i+p+1}z_{pq}b_{k+q+1}A^i \otimes B^k = \sum_{p,q} z_{pq}A_p \otimes B_q.$$

In this way we are led to our next result.

PROPOSITION 2.18. Let $Z = [z_{ik}] \in \mathbb{C}^{m\times n}$ be a solution of (2.35). Then (2.32) is uniquely solvable for all C, and the solution is determined by

$$X = \sum_{i=0}^{m-1}\sum_{k=0}^{n-1} z_{ik}A_iCB_k. \tag{2.37}$$

Similarly, we have

PROPOSITION 2.18'. Let $Z = [z_{ik}] \in \mathbb{C}^{m\times n}$ be a solution of (2.35'). Then (2.33) is uniquely solvable for all C, and the solution is determined by (2.37).

The initial point of further considerations of the equations (2.35), (2.35') is the following obvious fact.

REMARK 2.7. The solutions of (2.35) are H-matrices, the solutions of (2.35') are T-matrices.

We set

$$Z := [z_{i+k}] = H_{mn}(z) \tag{2.38}$$

for a solution of (2.35) and

$$Z := [z_{i-k}] = T_{mn}(z) \tag{2.38'}$$

for a solution of (2.35').

PROPOSITION 2.19. Suppose the resultant equation

$$\mathrm{Res}(a,b)z = e_{m+n-1} \tag{2.39}$$

is solvable and $z = (z_k)_0^{m+n-1}$. Then (2.35) is solvable and Z defined by (2.38) is a solution of this equation. Vice versa, if $Z = [z_{i+k}]$ is a solution of (2.35) and

$$z_{m+n-1} := -a_0 z_{n-1} - \dots - a_{m-1} z_{m+n-2}$$

then $z := (z_i)_0^{m+n-1}$ is a solution of (2.39).

The proof is an easy calculation.
Analogously, the following assertion holds.

PROPOSITION 2.19'. Suppose the resultant equation

$$\mathrm{Res}(a,\hat{b})z = -e_{m+n-1} \tag{2.40}$$

is solvable, $z := (z_k)_{1-n}^{m} \in \mathbb{C}^{m+n}$, then (2.35') is solvable and Z defined by (2.38') is a solution of (2.35'). If, vice versa, $Z = [z_{i-j}]$ is a solution of (2.35') and $z_m := -a_0 z_0 - \dots - a_{m-1} z_{m-1}$ then $z = (z_i)_{1-n}^{m}$ is a solution of (2.40).

To transform the latter properties into the Bezoutian language we have to show how the general case of different m and n can be reduced to the case $m = n$. Suppose, without loss of generality, $m > n$ and consider $b(\lambda)$ as an element of $\mathbb{C}^{m+1}(\lambda)$. Then the resultant matrix of $a(\lambda)$ and $b(\lambda)$ (for $p = 0$) is clearly a 2mx2m matrix. To distinguish between this matrix and the usual resultant matrix we shall use a subindex indicating the order of the matrix.
Then, obviously, the equation (2.39) together with the following definitions

$$z_k := -a_0 z_{k-m} - \dots - a_{m-1} z_{k-1} \quad (k=m+n,\dots,2m-1) \tag{2.41}$$

is equivalent to

$$\mathrm{Res}_{2m}(a,b)\tilde{z} = e_{2m-1}, \quad \tilde{z} := (z_i)_0^{2m-1}. \tag{2.42}$$

Now, let us consider for the general case instead of (2.40) the equation (2.42). Define still a number

$$z_{2m} := -a_0 z_m - \dots - a_{m-1} z_{2m-1} + 1.$$

Then the equation (2.42) is equivalent to

$$[a_0 a_1 \dots a_m] \begin{bmatrix} z_0 & z_1 & \dots & z_m \\ z_1 & & & \vdots \\ \vdots & & \cdot^{\cdot^{\cdot}} & \vdots \\ z_m & \dots & & z_{2m} \end{bmatrix} = [0 \dots 0 1],$$

$$[b_o \dots b_n 0 \dots 0] \begin{bmatrix} z_o & z_1 \dots & z_m \\ z_1 & & \\ \vdots & \cdot^{\cdot^{\cdot}} & \vdots \\ z_m & \dots & z_{2m} \end{bmatrix} = [0 \dots 01\gamma],$$

where γ is some complex number (in case $n < m$ we have $\gamma = 0$). The latter equations can be obviously written in the form

$$H_{m+1}(z)a = e_m,$$

$$H_{m+1}(z)(b - \gamma a) = e_{m-1},$$

where $z := (z_i)_o^{2m}$, $a := (a_i)_o^m$, $b := (b_i)_o^m$. Taking into account that $a_m \neq 0$, we can apply Theorem 1.4' and obtain that $H_m(z)$ is regular. Its inverse is just the Bezoutian of the polynomials $a(\lambda)$ and $b(\lambda) - \gamma a(\lambda)$ (see (1.48a)), which is obviously equal to the Bezoutian of $a(\lambda)$ and $b(\lambda)$.

Therefore, we have already proved the following

THEOREM 2.7. Suppose the spectra of the matrices $A \in \mathbb{C}^{m \times m}$ and $B \in \mathbb{C}^{n \times n}$ $(m \geq n)$ are disjoint and $Z = [z_{i+k}]_o^{m-1} \in \mathcal{H}(m)$ is the inverse of the Bezoutian of the characteristic polynomials of A and B. Then (2.32) is uniquely solvable and

$$X = \sum_{i=o}^{m-1} \sum_{k=o}^{n-1} z_{ik} A_i C B_k .$$

3. RECURSION FORMULAE AND INVERSION ALGORITHM FOR STRONGLY REGULAR T- AND H-MATRICES

In this section we realize the second step of fast T- and H-matrix inversion. We are interested in the deduction of algorithms for the solution of the fundamental equations with a low number of computations. This requires to investigate relations between the fundamental solutions for $T_n(a)$ and $H_n(b)$ for adjoining n. Let us remark that in this section inversion algorithms are obtained only for the case of strongly regular T- and H-matrices. The general case will be considered in Section 6.

Throughout this section we make use of the following designation. If $u \in \mathbb{C}^n$ then $\underline{u}$ denotes

$$\underline{u} := \begin{bmatrix} u \\ -1 \end{bmatrix} \in \mathbb{C}^{n+1}.$$

3.1. Recursion formulae - Toeplitz case. Suppose $a = (a_i)_{-N}^{N-1}$. Consider the fundamental solutions $x^n = (x_i^n)_o^{n-1}$, $y^n = (y_i^n)_o^{n-1}$, $z^n = (z_n^i)_o^{n-1}$ of the equations

$$T_n(a)x^n = e_o, \tag{3.1}$$

$$T_n(a)y^n = e_{n-1}, \tag{3.2}$$

$$T_n(a)z^n = g^n \ , \quad g^n := (a_{i-n})_o^{n-1} \tag{3.3}$$

for various $n \leq N$.

PROPOSITION 3.1. Suppose $T_n(a)$ and $T_{n+1}(a)$ are regular. Then $\varepsilon_n := [a_n, a_{n-1} \dots a_o]\underline{z}^n \neq 0$ and

$$x^{n+1} = \begin{bmatrix} x^n \\ 0 \end{bmatrix} - \frac{\alpha_n}{\varepsilon_n} \underline{z}^n \ , \tag{3.4}$$

$$z^{n+1} = \begin{bmatrix} 0 \\ z^n \end{bmatrix} - \beta_n \ x^{n+1} \ , \tag{3.5}$$

where $\alpha_n := [a_n, a_{n-1} \dots a_1]x^n$, $\beta_n := [a_{-1}, a_{-2} \dots a_{-n-1}]\underline{z}^n$.

PROOF. Obviously,

$$T_{n+1}(a)\begin{bmatrix} x^n \\ 0 \end{bmatrix} = e_o + \alpha_n e_n, \tag{3.6}$$

$$T_{n+1}(a)\underline{z}^n = \varepsilon_n e_n.$$

Since $T_{n+1}(a)$ is assumed to be regular, we have $\varepsilon_n \neq 0$. From these equalities immediately follows (3.4),(3.5). ≡

We have the following relation between the fundamental solutions z^n and y^{n+1}:

$$y^{n+1} = \frac{1}{\varepsilon_n} \underline{z}^n \ . \tag{3.7}$$

From this relation and Prop. 3.1 one obtains

PROPOSITION 3.2. Suppose $T_m(a)$ is regular for $m = n-1, n, n+1$ and denote $\gamma_n := [a_{-1}, a_{-2} \dots a_{-n}]y^n$. Then $\alpha_n\gamma_n \neq 1$ and

$$x^{n+1} = \frac{1}{1-\alpha_n\gamma_n} \left(\begin{bmatrix} x^n \\ 0 \end{bmatrix} - \alpha_n \begin{bmatrix} 0 \\ y^n \end{bmatrix}\right), \tag{3.8}$$

$$y^{n+1} = \frac{1}{1-\alpha_n\gamma_n} \left(\begin{bmatrix} 0 \\ y^n \end{bmatrix} - \gamma_n \begin{bmatrix} x^n \\ 0 \end{bmatrix}\right). \tag{3.9}$$

PROOF. Besides (3.6) we have

$$T_{n+1}(a) \begin{bmatrix} 0 \\ y^n \end{bmatrix} = \gamma_n e_o + e_n,$$

and (3.8), (3.9) follow immediately, provided that $\alpha_n\gamma_n \neq 1$. We prove the latter inequality. Suppose $\alpha_n\gamma_n = 1$, then for

$$u := \begin{bmatrix} x^n \\ 0 \end{bmatrix} - \alpha_n \begin{bmatrix} 0 \\ y^n \end{bmatrix}$$

$T_{n+1}(a)u = 0$ holds. Since, by assumption, $T_{n-1}(a)$ is regular, we have

$x_o^n \neq 0$, hence $u \neq 0$. This is a contradiction to the regularity of $T_{n+1}(a)$ and completes the proof. $\equiv$

Analogously to Prop. 3.2 recursion formulae for the other fundamental solutions can be deduced.

We are mainly interested, of course, in the recursion $n \to n+1$, but sometimes we also need the recursion $n+1 \to n$.

PROPOSITION 3.3. Suppose $T_m(a)$ is regular for $m = n, n+1$. Then $x_o^{n+1} \neq 0$ and

$$z_i^n = z_{i+1}^{n+1} - \frac{z_o^{n+1}}{x_o^{n+1}} x_{i+1}^{n+1} , \qquad (3.10)$$

$$(i=0,\dots,n-1)$$

$$x_i^n = x_i^{n+1} + x_n^{n+1} z_{i+1}^{n+1} - \frac{x_n^{n+1} z_o^{n+1}}{x_o^{n+1}} x_{i+1}^{n+1} . \qquad (3.11)$$

PROOF. From (3.5) we obtain

$$\left[{0 \atop z^n}\right] = z^{n+1} + \beta_n x^{n+1}. \qquad (3.12)$$

In particular, we have $0 = z_o^{n+1} + \beta_n x_o^{n+1}$, and (3.10) is obtained immediately. Furthermore, (3.4) implies

$$U_{n+1} x^{n+1} = \left[{0 \atop x^n}\right] - \frac{\alpha_n}{\varepsilon_n} \left[{0 \atop z^n}\right] \quad \text{and} \quad x_n^{n+1} = \frac{\alpha_n}{\varepsilon_n} .$$

In view of (3.12), we get

$$\left[{0 \atop x^n}\right] = U_{n+1} x^{n+1} + x_n^{n+1}(z^{n+1} + \beta_n x^{n+1}),$$

which is (3.11), and the proof is complete. $\equiv$

PROPOSITION 3.4. Suppose $T_m(a)$ is regular for $m = n, n+1$. Then $x_o^{n+1} = y_n^{n+1} \neq 0$

$$x_i^n = x_i^{n+1} - \frac{x_n^{n+1}}{x_o^{n+1}} y_i^{n+1} , \qquad (3.13)$$

$$(i=0,\dots,n-1)$$

$$y_i^n = y_{i+1}^{n+1} - \frac{y_o^{n+1}}{x_o^{n+1}} x_{i+1}^{n+1} . \qquad (3.14)$$

REMARK 3.1. All formulae noted above can be written in "polynomial" form:

$$x^{n+1}(\lambda) = x^n(\lambda) - \frac{\alpha_n}{\varepsilon_n} \underline{z}^n(\lambda), \qquad (3.4')$$

$$z^{n+1}(\lambda) = \lambda z^n(\lambda) - \beta_n x^{n+1}(\lambda), \qquad (3.5')$$

$$x^{n+1}(\lambda) = \frac{1}{1-\alpha_n\gamma_n}(x^n(\lambda) - \alpha_n \lambda y^n(\lambda)), \qquad (3.8')$$

$$y^{n+1}(\lambda) = \frac{1}{1-\alpha_n\gamma_n}(\lambda y^n(\lambda) - \gamma_n x^n(\lambda)), \tag{3.9'}$$

$$x^n(\lambda) = x^{n+1}(\lambda) - \frac{x_n^{n+1}}{x_0^{n+1}} y^{n+1}(\lambda), \tag{3.13'}$$

$$\lambda y^n(\lambda) = y^{n+1}(\lambda) - \frac{y_0^{n+1}}{x_0^{n+1}} x^{n+1}(\lambda). \tag{3.14'}$$

3.2. Recursion formulae - Hankel case. We deduce now the analogous recursion formulae for H-matrices and shall see that there are some essential differences between T- and H-matrices. Suppose $b = (b_i)_0^{2N-1}$ and consider the fundamental equations for $H_n(b)$ für various $n \leq N$.

$$H_n(b)y^n = e_{n-1}, \tag{3.15}$$

$$H_n(b)w^n = h^n, \quad h^n = (b_{n+i})_0^{n-1}. \tag{3.16}$$

PROPOSITION 3.5. Suppose $H_m(b)$ is regular for $m = n, n+1$. Then $\varepsilon_n := [b_n, b_{n+1} \dots b_{2n}]\underline{w}^n \neq 0$ and

$$y^{n+1} = \frac{1}{\varepsilon_n}\underline{w}^n, \tag{3.17}$$

$$w^{n+1} = \begin{bmatrix} 0 \\ w^n \end{bmatrix} - \varepsilon_n \begin{bmatrix} y^n \\ 0 \end{bmatrix} - (\beta_n - \gamma_n)\varepsilon_n y^{n+1}, \tag{3.18}$$

where $\beta_n := \frac{1}{\varepsilon_n}[b_{n+1} \dots b_{2n+1}]\underline{w}^n$, $\gamma_n := [b_n \dots b_{2n-1}]y^n$.

PROOF. We have

$$H_{n+1}(b)\underline{w}^n = \varepsilon_n e_n.$$

Since $H_{n+1}(b)$ is assumed to be regular we obtain $\varepsilon_n \neq 0$ and (3.17). In order to verify (3.18) we observe

$$H_{n+1}(b)\begin{bmatrix} y^n \\ 0 \end{bmatrix} = e_{n-1} + \gamma_n e_n,$$

$$H_{n+1}(b)\begin{bmatrix} 0 \\ w^n \end{bmatrix} = h^{n+1} + \varepsilon_n e_{n-1} + \beta_n \varepsilon_n e_n.$$

From these equalities (3.18) is easily deduced. $\equiv$

REMARK 3.2. Obviously, $\gamma_n = \beta_{n-1}$.

From Prop. 3.5 follows an $n-1, n \to n+1$ recursion, which is typical for H-matrices.

COROLLARY 3.1. Suppose $H_m(b)$ is regular for $m = n-1, n, n+1$. Then $\varepsilon_{n-1}\varepsilon_n \neq 0$ and

$$\underline{w}^{n+1} = \begin{bmatrix} 0 \\ \underline{w}^n \end{bmatrix} - \frac{\varepsilon_n}{\varepsilon_{n-1}} \begin{bmatrix} \underline{w}^{n-1} \\ 0 \\ 0 \end{bmatrix} + (\beta_{n-1} - \beta_n)\begin{bmatrix} \underline{w}^n \\ 0 \end{bmatrix}. \tag{3.19}$$

Furthermore, by Remark 3.2 and (3.17), one obtains

$$\varepsilon_{n+1}y^{n+2} = \varepsilon_n(\begin{bmatrix}0\\y^{n+1}\end{bmatrix} - \begin{bmatrix}y^n\\0\\0\end{bmatrix} + (\gamma_n-\gamma_{n+1})\begin{bmatrix}y^{n+1}\\0\end{bmatrix}). \tag{3.20}$$

PROPOSITION 3.6. Suppose $H_m(b)$ is regular for $m = n-1,n,n+1$ and denote $\xi_m := y^m_{m-1}$, $\zeta_m := y^m_{m-2}$. Then $\xi_n\xi_{n+1} \neq 0$ and

$$\begin{bmatrix}y^{n-1}\\0\\0\end{bmatrix} = -\frac{\xi_n}{\xi_{n+1}}\, y^{n+1} + \begin{bmatrix}0\\y^n\end{bmatrix} + \left(\frac{\zeta_{n+1}}{\xi_{n+1}} - \frac{\zeta_n}{\xi_n}\right)\begin{bmatrix}y^n\\0\end{bmatrix}. \tag{3.21}$$

PROOF. By relation (3.20) we have

$$y^{n-1}(\lambda) = ay^{n+1}(\lambda) + \lambda y^n(\lambda) + by^n(\lambda)$$

for some constants a and b. Comparing the highest order coefficients we obtain

$$0 = a\xi_{n+1} + \xi_n,$$

hence

$$a = -\frac{\xi_n}{\xi_{n+1}}.$$

Comparing the last but one coefficients we obtain

$$0 = a\zeta_{n+1} + \zeta_n + b\xi_n.$$

Consequently,

$$b = \frac{\zeta_{n+1}}{\xi_{n+1}} - \frac{\zeta_n}{\xi_n},$$

and the proposition is proved. ≡

3.3. Inversion algorithm for strongly regular T- and H-matrices.

DEFINITION 3.1. A matrix $A = [a_{ij}]_0^{N-1}$ is called strongly regular if each of its principal sections $A_n = [a_{ij}]_0^n$ $(n < N)$ is regular.

We note that the class of strongly regular matrices obviously includes the positive definite matrices.

Now we are going to describe how the fundamental solutions for strongly regular T- and H-matrices can be constructed with computational complexity $O(N^2)$ and storage requirements $O(N)$.

Let $T_N(a)$, $H_N(b)$ be strongly regular. Then, by definition, the sections $T_n(a)$, $H_n(b)$ fulfil the assumptions of section 3.1, 3.2 for any $n \leq N$. Basing upon this fact the following pictures shall illustrate the lapse of the inversion algorithms. Here the subscripts of the arrows are the numbers of formulae, pictures and the like used for the corresponding step.

PICTURE 3.1. Given a strongly regular T-matrix $T_N(a)$-determine its inverse $T_N(a)^{-1}$.

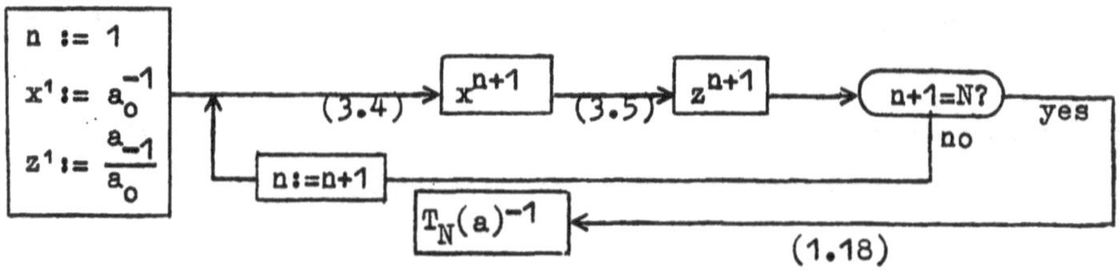

PICTURE 3.2. Given a strongly regular H-matrix $H_N(b)$ - determine its inverse $H_N(b)^{-1}$.

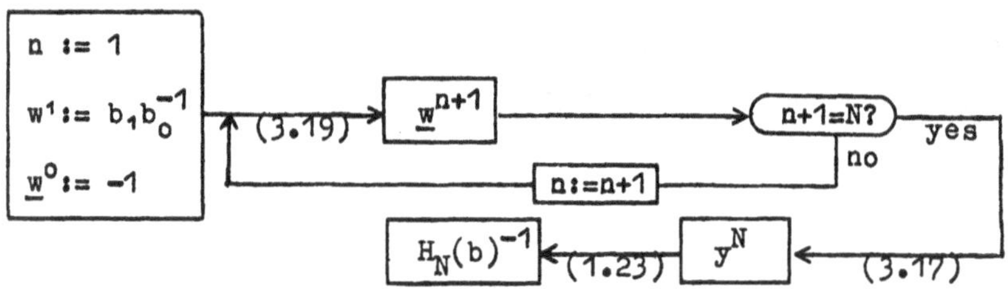

REMARK 3.3. One can evaluate that the algorithms of Pictures 3.1, 3.2 require $\frac{9}{2} N^2 + O(N)$ multiplications in the Toeplitz case and $4N^2 + O(N)$ in the Hankel case. (The inversion of Hermitian or symmetric T-matrices requires $\frac{5}{2} N^2 + O(N)$ multiplications.) They are more effective than, for example, the Gaussian one if $N > 5$.

3.4. Fast solution of Toeplitz and Hankel systems of equations.

Sometimes it is not necessary to determine the inverse matrix, but only to determine the solution of one special equation. This can also be done using recursion formulae.
Suppose we have to solve a Toeplitz system

$$T_N(a)\varphi = f, \tag{3.22}$$

where $T_N(a)$ is assumed to be strongly regular and $f = (f_i)_0^{N-1}$ is a given vector. Besides (3.22) we consider the fitted equations

$$T_n(a)\varphi^n = f^n, \quad f^n = (f_i)_0^{n-1}.$$

LEMMA 3.1. The following recursion holds

$$\varphi^{n+1} = \begin{bmatrix}\varphi^n \\ 0\end{bmatrix} - \frac{\eta_n}{\varepsilon_n} \underline{z}^n , \tag{3.23}$$

where $\eta_n := [a_n \dots a_1]\varphi^n - f_n$ and ε_n is as in Prop. 3.1.

Formula (3.23) is proved by applying $T_{n+1}(a)$. ≡
Let us mention that the formulae (3.4) and (3.7) are special cases of (3.23).
Basing on (3.23), (3.4) and (3.5) a recursion algorithm for the solution of (3.22) can be obtained. We illustrate it again in a picture.

PICTURE 3.3. Given a Toeplitz system (3.22), where $T_N(a)$ is strongly regular - determine $\varphi = \varphi^N$.

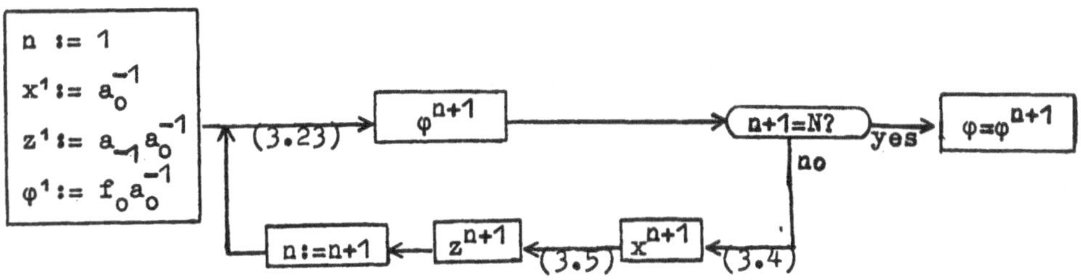

Analogously, it is possible to achieve solutions of H-systems

$$H_N(b)\varphi = f \tag{3.24}$$

if $H_N(b)$ is strongly regular. Consider again the fitted equations

$$H_n(b)\varphi^n = f^n \quad (n = 1,\dots,N).$$

LEMMA 3.1'. The following recursion holds

$$\varphi^{n+1} = \begin{bmatrix}\varphi^n\\0\end{bmatrix} - \frac{\eta_n}{\varepsilon_n}\,\underline{w}^n, \tag{3.25}$$

where $\eta_n := [b_n \dots b_{2n-1}]\varphi^n - f_n$ and ε_n is as in Prop. 3.5.
Formula (3.25) is proved by applying $H_{n+1}(b)$. ≡
We illustrate the solution of (3.24).

PICTURE 3.4. Given a Hankel system (3.24), where $H_N(b)$ is strongly regular - determine $\varphi = \varphi^N$.

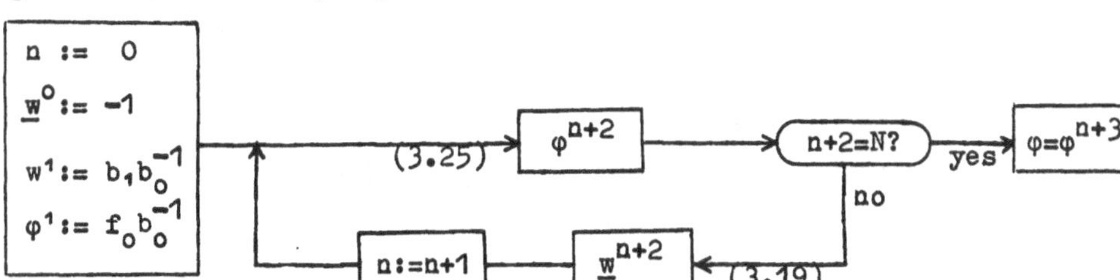

3.5. Inversion of perturbed T-matrices. Now we are interested to establish a fast algorithm for inversion of matrices A of the form

$$A = T_N(a) - K, \tag{3.26}$$

where $T_N(a)$ is a regular T-matrix and K is a matrix of rank $r \ll N$. Clearly, K can be represented in the form

$$K = G\Gamma$$

with $$G \in \mathbb{C}^{N\times r}, \quad \Gamma \in \mathbb{C}^{r\times N}. \tag{3.27}$$

LEMMA 3.2. Let A be a regular matrix of the form (3.26) with regular $T_N(a)$ and K defined by (3.27). Then the matrix

$$R = I_r - \Gamma T_N(a)^{-1}G$$

is regular, and the inverse of A can be written as

$$A^{-1} = T_N(a)^{-1} + T_N(a)^{-1}GR^{-1}\Gamma T_N(a)^{-1}. \tag{3.28}$$

PROOF. During the proof we write for short T instead of $T_N(a)$. The regularity of R immediately follows from

$$\Gamma T^{-1}A = \Gamma(I_N - T^{-1}G\Gamma) = R\Gamma,$$

since Γ has maximal rank.

Now we make some straightforward calculations to show that $A^{-1}A = I_N$:

$$A^{-1}A = (T^{-1} + T^{-1}GR^{-1}\Gamma T^{-1})(T - G\Gamma)$$

$$= I_N + T^{-1}GR^{-1}\Gamma - T^{-1}G\Gamma - T^{-1}GR^{-1}\Gamma T^{-1}G\Gamma.$$

Denoting by W the matrix $\Gamma T^{-1}G$ we obtain for the latter summand

$$- T^{-1}G(I_r - W)^{-1}W\Gamma = T^{-1}G(I_r - (I_r - W)^{-1})\Gamma.$$

This completes the proof. ≡

Now F, Φ denote the solutions of the matrix equations

$$T_N(a)F = G, \quad \Phi T_N(a) = \Gamma, \tag{3.29}$$

and we write equality (3.28) in the form

$$A^{-1} = T_N(a)^{-1} + F(I_r - \Gamma F)^{-1}\Phi. \tag{3.30}$$

In this way we are led to the following fast $O(N^2)$ - complexity algorithm.

PICTURE 3.5. Given regular A of the form (3.26), where K defined by (3.27) and $T_N(a)$ is strongly regular - determine the inverse of A.

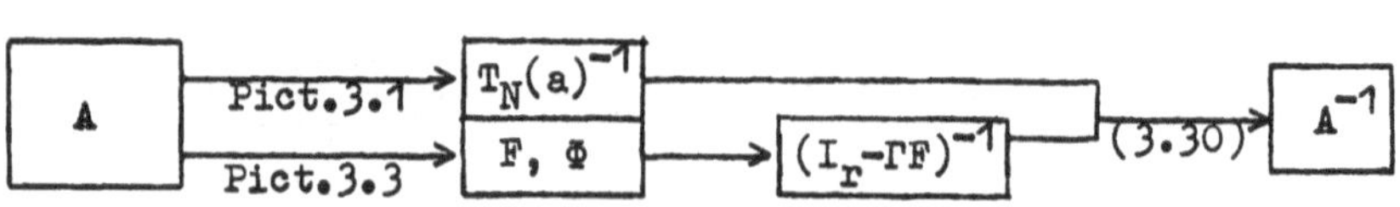

3.6. Lower-upper and upper-lower factorization. The recursion formulae can be used for the evaluation of the lower-upper factorization of T- and H-matrices and their inverses. Comparing the factorization with the inversion formulae of Section 1 one obtains the Christoffel-Darboux

formulae, well-known from the theory of orthogonal polynomials.
Let us recapitulate some concepts and properties closed to the factorization problem appearing in the linear algebra. We denote by $\mathcal{L}^n_-$ (respectively $\mathcal{L}^n_+$) the class of all nxn complex lower (upper) triangular matrices with entries equal to one on the main diagonal and by $\mathcal{D}^n$ the class of all nxn diagonal matrices. A lower-upper-factorization (for short "LU-factorization") of a regular matrix A is, by definition, a representation of the form

$$A = A_- D A_+, \tag{3.31}$$

where $A_\pm \in \mathcal{L}^n_\pm$ and $D \in \mathcal{D}^n$. Similarly an upper-lower factorization ("UL-factorization")

$$A = A_+ D A_- \tag{3.32}$$

is defined.
We collect some simple properties of the factorization:

(a) The LU- and UL-factorization is (provided that it exists) always unique.

(b) If A admits an LU-factorization (3.31) then the transposed A^T also admits an LU-factorization $A^T = B_- D B_+$, where $B_+ = A_-^T$, $B_- = A_+^T$. In particular, if A is symmetric then $A_- = A_+^T$.

(c) A admits an LU-factorization if and only if A^{-1} admits a UL-factorization.

(d) A admits an LU-factorization if and only if $J_n A J_n$ admits a UL-factorization.

(e) Suppose $A = [a_{ik}]_0^{n-1}$ admits an LU-factorization (3.31), then all principal sections $A_m = [a_{ik}]_0^{m-1}$ $(m \leq n)$ admit an LU-factorization

$$A_m = (A_-)_m D_m (A_+)_m \quad (m = 1,\dots,n).$$

In particular, matrices admitting an LU-factorization are strongly regular.

Now we shall show the converse of the last assertion, i.e. any strongly regular matrix admits an LU-factorization. Before doing this let us introduce some notation. For a given sequence of vectors $Z = \{z^1,\dots,z^n\}$ with $z^m = (z_i^m)_0^{m-1} \in \mathbb{C}^m$ we denote by $L_\pm(Z) = L_\pm(z^1,\dots,z^n)$ the matrices

$$L_+(Z) := \begin{bmatrix} z_0^1 & z_0^2 & \cdots & z_0^n \\ & z_1^2 & \cdots & z_1^n \\ & & \ddots & \vdots \\ & & & z_{n-1}^n \end{bmatrix}, \quad L_-(Z) := \begin{bmatrix} z_0^n & & & \\ z_1^n & z_0^{n-1} & & \\ \vdots & & \ddots & \\ z_{n-1}^n & z_{n-2}^{n-1} & \cdots & z_0^1 \end{bmatrix}.$$

Suppose now A is strongly regular and $Y = \{y^1,\dots,y^n\}$, where $y^k = (y_i^k)_0^{k-1}$ $(k = 1,\dots,n)$ are the solutions of the equations

$$A_k y^k = e_{k-1} \qquad (k = 1,\dots,n) . \tag{3.33}$$

We observe that the matrix $AL_+(Y)$ belongs to $\mathcal{L}^n_-$. Therefore, A admits a factorization (3.31) with

$$A_- := AL_+(Y),\ D := \operatorname{diag}(\frac{1}{y_0^1}\dots\frac{1}{y_{n-1}^n}),\quad A_+ := D^{-1}L_+(Y)^{-1}.$$

We remark that we have also

$$A_-^T = D^{-1}L_+(Y')^{-1}, \tag{3.34}$$

where $Y' = \{y'^1,\dots,y'^n\}$ and y'^k are the solutions of

$$A_k^T y'^k = e_{k-1} \qquad (k = 1,\dots,n). \tag{3.35}$$

Similarly, in order $A = [a_{ij}]_0^{n-1}$ admit a UL-factorization (3.32) it is necessary and sufficient that J_nAJ_n is strongly regular, which is equivalent to the regularity of all sections

$$A_{(k)} := [a_{ij}]_k^{n-1} \qquad (k = 0,\dots,n-1).$$

Thus, for determining the UL-factorization one has to solve the equations

$$A_{(k)}x^k = e_0 \tag{3.36}$$

and $\qquad (k = 1,\dots,n)$

$$A_{(k)}^T x'^k = e_0. \tag{3.37}$$

The factorization problem can be formulated in terms of generating functions. Provided that there is an LU-factorization (3.31) with $A_- = L_-(a_-^1 \dots a_-^n)$, $A_+ = L_-(a_+^1 \dots a_+^n)^T$ and $D = \operatorname{diag}(d_1\dots d_n)$ we obtain

$$A(\lambda,\mu) = \sum_{k=1}^n a_-^k(\lambda)d_{n+1-k}a_+^k(\mu)(\lambda\mu)^{n-k}. \tag{3.38}$$

Analogously, if $A = A_+DA_-$ is a UL-factorization with $A_+ = L_+(a_+^1\dots a_+^n)$, $A_- = L_+(a_-^1\dots a_-^n)^T$, $D = \operatorname{diag}(d_1\dots d_n)$, then

$$A(\lambda,\mu) = \sum_{k=1}^n a_+^k(\lambda)d_k a_-^k(\mu). \tag{3.39}$$

Consider now the special case that $A = H_n(b)$ is a strongly regular H-matrix. Since H-matrices are symmetric we have $Y' = Y$.
Clearly, from the LU-factorization of $H_n(b)$ follows a UL-factorization of the Bezoutian $H_n(b)^{-1}$:

$$H_n(b)^{-1} = L_+(Y)DL_+^T(Y),$$

where $D = \operatorname{diag}((y_0^1)^{-1},\dots,(y_{n-1}^n)^{-1})$. Comparing this with inversion formula (1.48b) we obtain, in virtue of (3.39), a Christoffel-Darboux formula presented in the following proposition.

PROPOSITION 3.7. Let $H_n(b)$ be a strongly regular H-matrix and y^k the solutions of the equations (3.33). Then

$$\sum_{k=1}^{n-1} \frac{1}{y_{k-1}^{k}} y^k(\lambda)y^k(\mu) = \frac{1}{y_{n-1}^{n}} \frac{y^n(\lambda)y^{n-1}(\mu)-y^{n-1}(\lambda)y^n(\mu)}{\lambda - \mu} .$$

We showed above that the columns of the factorization factor $L_+(Y)$ of the inverse of a strongly regular H-matrix fulfil the recurrent formula (3.20). It turns out that this property is characteristic for such matrices.

PROPOSITION 3.8. Suppose $A \in \mathbb{C}^{n\times n}$ is defined by $A = A_+DA_-$, where $A_+ = A_-^T = L_+(y^1,\dots,y^n)$, $D = \mathrm{diag}(d_1,\dots,d_n)$, and y^k fulfil the recurrent formula

$$\lambda d_k y^k(\lambda) = d_{k+1}y^{k+1}(\lambda) + a_k y^k(\lambda) + d_k y^{k-1}(\lambda) \;^{1)} \tag{3.40}$$

for some numbers d_{n+1}, a_k $(k = 1,\dots,n)$. Then A is the Bezoutian

$$A = \mathrm{Bez}(d_{n+1}y^{n+1},y^n). \tag{3.41}$$

PROOF. We have, due to (3.39),

$$A(\lambda,\mu) = \sum_{k=1}^{n} y^k(\lambda)d_k y^k(\mu),$$

which implies according to (3.40)

$$\begin{aligned}(\lambda-\mu)A(\lambda,\mu) &= \sum_{k=1}^{n} [(d_{k+1}y^{k+1}(\lambda)+a_k y^k(\lambda)+d_k y^{k-1}(\lambda))y^k(\mu) \\ &\quad - y^k(\lambda)(d_{k+1}y^{k+1}(\mu)+a_k y^k(\mu)+d_k y^{k-1}(\mu))] \\ &= d_{n+1}(y^{n+1}(\lambda)y^n(\mu) - y^n(\lambda)y^{n+1}(\lambda)).\end{aligned}$$

Hence (3.41) holds, and the proposition is proved. ≡

COROLLARY 3.2. A strongly regular matrix is Hankel if and only if its fundamental solutions (3.33) satisfy a recursion (3.40).

Let us shortly remark a connection of the H-matrix factorization to the theory of orthogonal polynomials. Suppose $\mu(t)$ is a real-valued function of bounded variation on $[\alpha,\beta]$ and $b = (b_k)_0^{2n-2}$,

$$b_k = \int_\alpha^\beta t^k d\mu(t).$$

Then we have for polynomials $p(\lambda)$, $q(\lambda) \in \mathbb{R}^n(\lambda)$

$$\langle p,q\rangle := \int_\alpha^\beta p(t)\overline{q}(t)d\mu(t) = q^* H_n(b)p.$$

In particular, provided that $H_n(b)$ is strongly regular, for the fundamental solutions y^k there is $\langle y^k, y^l\rangle = 0$ if $k \neq l$.

---

1) We put $y^0 := 0$.

The polynomials y^k can be normalized

$$\overset{o}{y}^k := \frac{1}{\sqrt{y^k_{k-1}}} y^k,$$

such that

$$\langle \overset{o}{y}^k, \overset{o}{y}^l \rangle = \delta_{kl},$$

i.e. $\overset{o}{y}^k(\lambda)$ is a system of orthonormal with respect to $\mu(t)$ polynomials. Let us still remark that the normed polynomials fulfil the recursion

$$\lambda \overset{o}{y}^k(\lambda) = d_k \overset{o}{y}^{k+1}(\lambda) + c_k \overset{o}{y}^k(\lambda) + d_{k-1} \overset{o}{y}^{k-1}(\lambda),$$

where

$$c_k := \gamma_k - \gamma_{k-1}, \quad d_k := \sqrt{\frac{\varepsilon_k}{\varepsilon_{k-1}}} \text{ .}^{1)}$$

Consider now strongly regular T-matrices $A = T_n(a)$. Since for T-matrices we have $A_k = A_{(k)}$ the matrix A is strongly regular if and only if $J_n A J_n$ is strongly regular. Moreover, we have simultaneously an LU- and a UL-factorization of $T_n(a)^{-1}$

$$T_n(a)^{-1} = L_-(X) D L_-(\hat{Y})^T = L_+(Y) \hat{D} L_+(\hat{X})^T,$$

where

$$X = \{x^1, \ldots, x^n\}, \qquad Y = \{y^1, \ldots, y^n\},$$

$$\hat{X} = \{\hat{x}^1, \ldots, \hat{x}^n\}, \qquad \hat{Y} = \{\hat{y}^1, \ldots, \hat{y}^n\}, \quad \hat{y}^k = J_k y^k$$

$$D = \mathrm{diag}\left(\frac{1}{x_0^n}, \ldots, \frac{1}{x_0^1}\right), \quad \hat{D} = J_n D J_n.$$

A comparison with the inversion formulae leads to

PROPOSITION 3.9. For a strongly regular T-matrix there is

$$\frac{1}{x_0^n} \frac{x^n(\lambda)\hat{y}^n(\mu) - y^n(\lambda)\mu\hat{x}^n(\mu)}{1 - \lambda\mu} = \sum_{k=1}^{n} (\lambda\mu)^{n-k} \frac{x^k(\lambda)\hat{y}^k(\mu)}{x_0^k}$$

$$= \sum_{k=1}^{n} \frac{y^k(\lambda)\hat{x}^k(\mu)}{x_0^k},$$

$$\frac{1}{x_0^n} \frac{x^n(\lambda)\hat{y}^n(\mu) - y^n(\lambda)\hat{x}^n(\mu)}{1 - \lambda\mu} = \sum_{k=1}^{n-1} (\lambda\mu)^{n-1-k} \frac{x^k(\lambda)\hat{y}^k(\mu)}{x_0^k}$$

$$= \sum_{k=1}^{n-1} \frac{y^k(\lambda)\hat{x}^k(\mu)}{x_0^k}.$$

We state a T-matrix analogue of Cor. 3.2.

---

1) For the definition of γ_k, ε_k see Prop. 3.5.

PROPOSITION 3.10. Suppose the matrices $A \in \mathbb{C}^{n\times n}$ and J_nAJ_n are strongly regular and the solutions of

$$A_kx^k = e_o, \quad A_ky^k = e_{k-1}$$

fulfil the recursion (3.13), (3.14). Then A is Toeplitz.

PROOF. By assumption A^{-1} admits an LU- and a UL-factorization. That means, for some polynomials $u^k(\lambda)$, $v^k(\lambda) \in \mathbb{C}^k(\lambda)$ we have

$$\begin{aligned} A^{-1}(\lambda,\mu) &= \sum_{k=1}^{n} y^k(\lambda)d_ku^k(\mu) \\ &= \sum_{k=1}^{n} (\lambda\mu)^{n-k}x^k(\lambda)\delta_{n+1-k}v^k(\mu), \end{aligned}$$

where $d_k := (y^k_{k-1})^{-1}$, $\delta_k := (x^k_o)^{-1}$.

One can conclude from (3.13), (3.14), by induction, that $d_k = \delta_k$. The proof that A is Toeplitz is made by induction. Clearly, the first section A_1 is Toeplitz. Suppose that A_{m-1} is Toeplitz and prove that then A_m is Toeplitz, too. By assumption, we have $u^k = \hat{x}^k$ and $v^k = \hat{y}^k$ for $k < m$. Therefore, we have, in virtue of Prop. 3.9,

$$\begin{aligned} A_m^{-1}(\lambda,\mu) &= A_{m-1}^{-1}(\lambda,\mu) + y^m(\lambda)d_mu^m(\mu) \\ &= \lambda\mu\, A_{m-1}^{-1}(\lambda,\mu) + x^m(\lambda)d_mv^m(\mu). \end{aligned}$$

We obtain

$$(1-\lambda\mu)A_m^{-1}(\lambda,\mu) = x^m(\lambda)d_mv^m(\mu) - y^m(\lambda)d_mu^m(\mu),$$

hence, A_m^{-1} is a T-Bezoutian, which is A_m is Toeplitz, and the proposition is proved. ≡

Analogously to the connections of H-matrices with orthogonal polynomials on an interval T-matrices have connections to orthogonal polynomials on the unit circle. Suppose $\mu(\zeta)$ is a measure on $\mathbb{T}$ (real but not necessary positive) and $a = (a_k)^n_{1-n}$,

$$a_k = \bar{a}_{-k} = \int_{\mathbb{T}} \zeta^k d\mu.$$

Then we have

$$\langle p,q\rangle := \int p(\zeta)\overline{q(\zeta)}d\mu = q^*T_n(a)p$$

for $p,q \in \mathbb{C}^n$, and the polynomials $y^k(\zeta)$ are orthogonal with respect to the inner product $\langle\cdot,\cdot\rangle$.

3.7. Evaluation of the signature in the strongly regular case.

In order to solve root localization problems one has, as we saw in Section 2, often to determine the signature of H-, T-matrices or Bezoutians. This can be done using the recursion formulae presented

above, provided the matrix is strongly regular. [1]
Suppose $A = C^*DC$ [2] with $D = \mathrm{diag}(d_1,\dots,d_n)$ is the UL- or LU-factorization of the Hermitian matrix A, then, by Sylvester's theorem,

$$\mathrm{sgn}\, A = \sum_{k=1}^{n} \mathrm{sgn}\, d_k. \tag{3.42}$$

In particular, for strongly regular real H-matrices we have

$$\mathrm{sgn}\, H_n(b) = -\sum_{k=1}^{n} \mathrm{sgn}\, \varepsilon_k = \sum_{k=1}^{n} \mathrm{sgn}\, y_{k-1}^{k}, \tag{3.43}$$

where ε_k is defined in Prop. 3.5. Analogously, for strongly regular Hermitian T-matrices there is by (3.8)

$$\mathrm{sgn}\, T_n(a) = \sum_{k=1}^{n} \mathrm{sgn}\, x_0^k. \tag{3.44}$$

In a similar manner we evaluate the signature of Bezoutians. Suppose $B = \mathrm{Bez}(a,b) \in \mathcal{B}(n)$ is an Hermitian Bezoutian with $J_n B J_n$ strongly regular. Then the highest order coefficients a_n and b_n are non-zero, and we have

$$y^n(\lambda) = b_n\, a(\lambda) - a_n\, b(\lambda),$$
$$y^{n+1}(\lambda) = \alpha\, a(\lambda) - \beta\, b(\lambda),$$

where $\alpha,\beta \in \mathbb{C}$ such that $\alpha\, a_n + \beta\, b_n = -1$. We obtain

$$\mathrm{sgn}\, B = \sum_{k=1}^{n} y_{k-1}^{k},$$

where $y^{n-1},\dots,y^1$ can be determined by means of Prop. 3.6.

<u>REMARK 3.4.</u> It is easy to see that any Hermitian Bezoutian is the Bezoutian of two real polynomials. More in detail, if $B := \mathrm{Bez}(a,b)$ is Hermitian then

$$B = c\, \mathrm{Bez}(\mathrm{Re}\, a,\ \mathrm{Im}\, b) \tag{3.45}$$

for some constant c. Indeed, we have

$$B = \mathrm{Bez}(a_r,b_r) - \mathrm{Bez}(a_i,b_i) + i(\mathrm{Bez}(a_r,b_i) + \mathrm{Bez}(a_i,b_r)),$$

where $a_r := \mathrm{Re}\, a$, $b_r := \mathrm{Re}\, b$, $a_i := \mathrm{Im}\, a$, $b_i := \mathrm{Im}\, b$.
Consequently, since B is Hermitian

$$\mathrm{Bez}(a_r,b_i) = -\mathrm{Bez}(b_r,a_i).$$

Taking into account Cor. 2.2 we achieve (3.45).
Let us now consider strongly regular Hermitian T-Bezoutians $B_T := \mathrm{Bez}_T(a,b)$. Then, obviously,

---

[1] For the general case see Section 6.

[2] The star designates the Hermitian adjoined matrix.

$$x^n(\lambda) = b_n a(\lambda) - a_n b(\lambda),$$

$$y^n(\lambda) = \hat{x}^n(\lambda),$$

and we obtain

$$\operatorname{sgn} B_T = \sum_{k=1}^{n} x_o^k,$$

where $x^{n-1},\dots,x^1$ can be determined by means of Prop. 3.4.

3.8. Inversion of Bezoutians. For solving the Liapunov and Stein equations one has to determine the inverse of a Bezoutian (see Subsection 2.10). Suppose $B := \operatorname{Bez}(c,d) \in \mathcal{B}(n)$ is a Bezoutian such that $J_n B J_n$ is strongly regular. We put

$$y^{n+1}(\lambda) := -\frac{c(\lambda)}{c_n}, \quad y^n(\lambda) = d_n c(\lambda) - c_n d(\lambda)$$

and determine successively the vectors $y^{n-1},\dots,y^1$ by means of Prop. 3.6. We know the inverse of B is a strongly regular H-matrix $H_n(b)$, $b = (b_i)_o^{2n-2}$, with y^i as the solution of the equations

$$H_i(b)y^i = e_{i-1} \qquad (i = 1,\dots,n).$$

Therefore, the matrix $H_n(b)$ can be established successively by help of the following recursion formulae

$$b_o = \frac{1}{y_o^1}, \quad b_{2k-1} = -\frac{1}{y_k^{k+1}} \sum_{i=o}^{k-1} b_{k-1+i} y_i^{k+1},$$
$$b_{2k} = \frac{1}{y_k^{k+1}}\Big(1 - \sum_{i=o}^{k-1} b_{k+i} y_i^{k+1}\Big). \qquad (k = 1,\dots,n-1)$$

The computational complexity for Bezoutian inversion is, obviously, $O(n^2)$, analogously to the H-matrix inversion. But we have to remark that, different to the H-matrix inversion, the algorithm offered here has a storage requirement of $O(n^2)$.
Now let us consider strongly regular T-Bezoutians $B_T := \operatorname{Bez}_T(c,d)$. We put

$$x^n(\lambda) := c(\lambda)d_n - d(\lambda)c_n,$$
$$y^n(\lambda) := c(\lambda)d_o - d(\lambda)c_o,$$

and determine successively the vectors x^{n-i}, y^{n-i} $(i = 1,\dots,n-1)$ by Prop. 3.4. The matrix B_T^{-1} is a strongly regular T-matrix $T_n(a)$, $a = (a_i)_{-n+1}^{n-1}$, and can be established successively by

$$a_o = \frac{1}{x_o^1}, \quad a_{-k} = -\frac{1}{x_o^{k+1}} \sum_{i=o}^{k-1} a_{-i} y_i^{k+1},$$
$$a_k = -\frac{1}{x_o^{k+1}} \sum_{i=1}^{k} a_{k-i} x_i^{k+1}. \qquad (k = 1,\dots,n-1)$$

4. TRANSFORMATIONS OF T- AND H-MATRICES AND BEZOUTIANS

4.0. Two problems. To any pair of matrices $A \in \mathbb{C}^{m\times m}$, $B \in \mathbb{C}^{n\times n}$ we associate the transformation $A \otimes B, (A \otimes B)X := AXB$ $(X \in \mathbb{C}^{m\times n})$ (see 2.10). In this section we shall give an answer of the following questions.

PROBLEM 1. For which $A \in \mathbb{C}^{m\times m}$ and $B \in \mathbb{C}^{n\times n}$ $A \otimes B$ transforms
a) $\mathcal{H}(m,n)$ into $\mathcal{H}(m,n)$
b) $\mathcal{H}(m,n)$ into $\mathcal{T}(m,n)$
c) $\mathcal{T}(m,n)$ into $\mathcal{H}(m,n)$
d) $\mathcal{T}(m,n)$ into $\mathcal{T}(m,n)$, respectively.
In virtue of Prop. 0.1 all problems 1a - 1d are solved if one of them is solved.
The second problem deals with quadratic forms. To any Hermitian matrix $Z \in \mathbb{C}^{n\times n}$ we associate the quadratic form $Z(x,x) = x^*Zx$ $(x \in \mathbb{C}^n)$.

PROBLEM 2. For which transformations of coordinates
a) H-forms [1] go over into H-forms
b) H-forms into T-forms
c) T-forms into H-forms and
d) T-forms into T-forms.
Let us explain the connection between these two problems. Suppose $y = Cx$ is a transformation of coordinates satisfying 2.a. Then for any Hermitian (real) H-matrix $H_n \in \mathcal{H}(n)$, $H_n(Cx,Cx)$ is, by assumption, an H-form. That means C^*H_nC is an H-matrix. Clearly, then C^*H_nC is Hankel for all $H_n \in \mathcal{H}(n)$ (not necessarily Hermitian). Therefore $C^* \otimes C$ is a transformation satisfying 1a. Vice versa, if $C^* \otimes C$ is a transformation satisfying 1.a, then $y = Cx$ satisfies 2.a. Analogous connections hold for b, c and d.

4.1. Möbius matrices. To solve the two problems stated above we have to introduce a special class of matrices.
We denote by $GL(\mathbb{C}^n)$ the group of regular $n\times n$ matrices. To any matrix $\varphi = \begin{bmatrix}\alpha & \gamma\\ \beta & \delta\end{bmatrix} \in GL(\mathbb{C}^2)$ we associate the operator in $\mathbb{C}^n(\lambda)$

$$M_n(\varphi)x := x\left(\frac{\alpha\lambda + \beta}{\gamma\lambda + \delta}\right)(\gamma\lambda + \delta)^{n-1}. \tag{4.1}$$

The matrices corresponding to an operator of this form (with respect to the basis $\{\lambda^k\}$) will be called Möbius matrices. The class of Möbius matrices will be denoted by $\mathcal{M}_n$. It is easy to verify that

$$M_n(\varphi\psi) = M_n(\varphi)M_n(\psi). \tag{4.2}$$

[1] We say, for short "H-forms" instead of Hankel quadratic forms and "T-forms" instead of Toeplitz quadratic forms.

In particular, we have for $\varepsilon \in \mathbb{C}$

$$M_n(\varepsilon\varphi) = \varepsilon^{n-1}M_n(\varphi). \tag{4.3}$$

Hence, the correspondence $\varphi \to M_n(\varphi)$ is a group homomorphism of $GL(\mathbb{C}^2)$ into $GL(\mathbb{C}^n)$ [1]. Moreover, it is a monomorphism, since $M_n(\varphi)=I_n$ implies $\varphi = I_2$. In particular, $\mathcal{M}_n$ is a subgroup of $GL(\mathbb{C}^n)$.
The group $GL(\mathbb{C}^2)$ is generated by the matrices

$$a(\beta) := \begin{bmatrix}1 & 0\\ \beta & 1\end{bmatrix}, \quad c(\alpha) := \begin{bmatrix}\alpha & 0\\ 0 & 1\end{bmatrix}, \quad j := \begin{bmatrix}0 & 1\\ 1 & 0\end{bmatrix} = J_2.$$

Therefore, the subgroup $\mathcal{M}_n$ is generated by the operators

(i) $M_n(a(\beta))x(\lambda) = x(\lambda + \beta)$

(ii) $M_n(c(\alpha))x(\lambda) = x(\alpha\lambda)$

(iii) $M_n(j)x(\lambda) \quad = \hat{x}(\lambda) = x(\lambda^{-1})\lambda^{n-1}$.

Let us still note an explicit representation of the entries of a Möbius matrix.

PROPOSITION 4.1. Suppose $\varphi = \begin{bmatrix}\alpha & \gamma\\ \beta & \delta\end{bmatrix} \in GL(\mathbb{C}^2)$ and $M_n(\varphi) = [m_{ik}]_0^{n-1}$, then

$$m_{ik} = \sum_{p=\max(i+k-n+1,o)}^{\min(i,k)} \binom{k}{p}\binom{n-k-1}{i-p}\alpha^p\beta^{k-p}\gamma^{i-p}\delta^{n-i-k+p-1}. \tag{4.4}$$

The equality (4.4) is obtained from

$$M_n(\varphi)e_k(\lambda) = (\alpha\lambda+\beta)^k(\gamma\lambda+\delta)^{n-k-1} = \sum_{i=1}^{n-1} m_{ik}\lambda^i$$

by a straightforward calculation. Special cases of (4.4) are the relations

$$M_n(a(\beta)) = [\binom{k}{i}\beta^{k-i}]_0^{n-1} \quad (\beta \neq 0), \tag{4.4a}$$

$$M_n(c(\alpha)) = [\alpha^i\delta_{ik}]_0^{n-1}, \quad M_n(j) = J_n. \tag{4.4b}$$

4.2. Transformations of T- and H-matrices. Now we can formulate the solution of Problem 1.

THEOREM 4.1. Suppose $m,n > 1$. A non-zero transformation $A \otimes B$ satisfies 1.a-d if and only if there is a $\varphi \in GL(\mathbb{C}^2)$ such that $B = M_n(\varphi)$ and, respectively,

a) $A = M_m(\varphi)^T$, b) $A = M_m(\varphi j)^T$,

c) $A = M_m(j\varphi)^T$, d) $A = M_m(j\varphi j)^T$.

Furthermore, the following equalities hold

a) $M_m(\varphi)^T H_{mn}(b)M_n(\varphi) = H_{mn}(M_{m+n-1}(\varphi)^T b)$,

[1] Remember that we identify matrices with the corresponding operators.

b) $M_m(\varphi j)^T H_{mn}^{m-1}(\hat{a}) M_n(\varphi) = T_{mn}(M_{m+n-1}(\varphi)^T a)$,

c) $M_m(j\varphi)^T T_{mn}(a) M_n(\varphi) = H_{mn}^{m-1}(M_{m+n-1}(j\varphi)^T a)$,

d) $M_m(j\varphi j)^T T_{mn}(a) M_n(\varphi) = T_{mn}(M_{n+n-1}(\varphi)^T a)$.

PROOF. It is sufficient to prove the a)-part of the theorem. Suppose $B = M_n(\varphi)$, $A = M_m(\varphi)^T$, where $\varphi \in GL(\mathbb{C}^2)$. If $\varphi = j$ or $\varphi = c(\alpha)$, then, obviously, $A \otimes B$ transforms $\mathcal{H}(m,n)$ into $\mathcal{H}(m,n)$. Let us assume now $\varphi = a(\beta)$, $b = (b_i)_0^{m+n-2}$ and $G := [g_{ik}]_{0}^{m-1}{}_{0}^{n-1} := (A \otimes B)\, H_{mn}(b)$. Then we have because of (4.4a)

$$g_{ik} = \sum_{p=0}^{i} \sum_{q=0}^{k} \binom{i}{p}\binom{k}{q} \beta^{i+k-p-q} b_{p+q}$$

$$= \sum_{r=0}^{i+k} \sum_{p=0}^{i} \binom{i}{p}\binom{k}{r-p} \beta^{i+k-r} b_r$$

$$= \sum_{r=0}^{i+k} \binom{i+k}{r} \beta^{i+k-r} b_r \ .$$

Therefore, G is an H-matrix, i.e. $A \otimes B$ transforms $\mathcal{H}(m,n)$ into $\mathcal{H}(m,n)$. Since the transformations j, $c(\alpha)$ and $a(\beta)$ generate $\mathcal{M}_n$, the sufficiency part of the theorem is proved.

Now let us assume that $A \otimes B$ transforms $\mathcal{H}(m,n)$ into $\mathcal{H}(m,n)$. Denote $l(\lambda) := (\lambda^k)_0^{m+n-2}$ for $\lambda \in \mathbb{C}$ and $l(\infty) := e_{m+n-2}$, then $H_\lambda := H_{mn}(l(\lambda))$ is an H-matrix of rank 1. On the other side, as it is easily seen, any H-matrix of rank 1 has the form ϱH_λ ($\varrho \in \mathbb{C}$, $\lambda \in \overline{\mathbb{C}}$), provided that $m,n > 1$. By assumption $G_\lambda := (A \otimes B) H_\lambda$ is an H-matrix of rank 1. Therefore, there exist functions $\varrho(\lambda) : \overline{\mathbb{C}} \to \mathbb{C}$ and $\varphi(\lambda) : \overline{\mathbb{C}} \to \overline{\mathbb{C}}$ such that $G_\lambda = \varrho(\lambda) H_{\varphi(\lambda)}$. In case $\varrho(\lambda) \equiv 0$ $A \otimes B$ is the null transformation. This follows from the fact that the matrices H_λ form a complete system in the space $\mathcal{H}(m,n)$. Therefore $\varrho(\lambda) \not\equiv 0$. Furthermore, H_λ is a matrix polynomial of degree $m + n - 2$, thus G_λ must be such a polynomial, too. That means the functions $\varrho(\lambda)\varphi(\lambda)^k$ ($k = 0,\dots,m+n-2$) are polynomials of degree $\le m+n-2$. In particular, $\varrho(\lambda)$ is a polynomial and $\varphi(\lambda)$ is a rational function. Suppose $\varphi(\lambda) = p(\lambda)/q(\lambda)$, where $p(\lambda)$ and $q(\lambda)$ are disjoint. Since $\varrho(\lambda)\varphi(\lambda)^{m+n-2}$ is a polynomial, $\varrho(\lambda)$ is a multiple of $q(\lambda)^{m+n-2}$. Further, since $\varrho(\lambda)$ has at most degree $m+n-2$, $q(\lambda)$ must by linear. Suppose $q(\lambda) = \gamma\lambda + \delta$, where $\gamma \not= 0$. Then $\varrho(\lambda) = \varepsilon(\gamma\lambda + \delta)^{m+n-2}$ for some $\varepsilon \not= 0$. We have

$$\varrho(\lambda)\varphi(\lambda)^k = \varepsilon p(\lambda)^k (\gamma\lambda + \delta)^{m+n-2-k}.$$

From this one concludes that $p(\lambda)$ is linear. Suppose $p(\lambda) = \alpha\lambda + \beta$ we obtain

$$\varphi(\lambda) = \frac{\alpha\lambda + \beta}{\gamma\lambda + \delta}, \quad \varrho(\lambda) = \varepsilon(\gamma\lambda + \delta)^{m+n-2}. \tag{4.5}$$

Consider the case $\gamma = 0$. In this case $p(\lambda)^{m+n-2}\varrho(\lambda)$ is a polynomial of degree $\leq m+n-2$. Therefore, $p(\lambda)$ is linear and $\varrho(\lambda)$ is constant. With $p(\lambda) = \alpha\lambda + \beta$ we have here again $\varphi(\lambda)$ and $\varrho(\lambda)$ as in (4.5) for some $\varepsilon \in \mathbb{C}$.

Let θ be such that $\theta^{m+n-2} = \varepsilon$. We set

$$\varphi := \theta\begin{bmatrix}\alpha & \gamma\\ \beta & \delta\end{bmatrix}$$

and show that $B = M_n(\varphi)$ and $A = M_m(\varphi)^T$. For vectors $x \in \mathbb{C}^n$ and $y \in \mathbb{C}^m$ there is

$$\begin{aligned} y^T G_\lambda x &= \varrho(\lambda) y(\varphi(\lambda))^T x(\varphi(\lambda)) \\ &= \varepsilon(\gamma\lambda + \delta)^{m+n-2} y(\varphi(\lambda))^T x(\varphi(\lambda)) \\ &= y(\varphi(\lambda))^T(\theta\gamma\lambda + \theta\delta)^{m-1} x(\varphi(\lambda))(\theta\gamma\lambda + \theta\delta)^{n-1} \\ &= ((M_m(\varphi)y)(\lambda))^T(M_n(\varphi)x)(\lambda). \end{aligned}$$

This implies

$$y^T G_\lambda x = y^T M_m(\varphi)^T H_\lambda M_n(\varphi)x.$$

Consequently, $A = M_m(\varphi)^T$ and $B = M_n(\varphi)$. It remains to prove the equality a). For $\varphi = j$, $\varphi = c(\alpha)$ it is obvious, and for $\varphi = a(\beta)$ we have to make some elementary calculations.

The other equalities immediately follow from a) and the well-known relation

$$J_m H_{mn}^{m-1}(\hat{a}) = T_{mn}(a).$$

This completes the proof of the theorem. ≡

<u>REMARK 4.1.</u> We have proved that any non-zero transformation $A \otimes B$ acting within $\mathcal{H}(m,n)$ or $\mathcal{T}(m,n)$ is an isomorphism.

<u>4.3. Transformations of T- and H-forms.</u> We are going now to present the solution of Problem 2. Taking into account Theorem 4.1 and the remarks at the beginning of this section this problem is reduced to the question whether

$$\begin{array}{llll} \text{a)} & M_n(\varphi)^T = M_n(\varphi)^* & \text{b)} & M_n(\varphi j)^T = M_n(\varphi)^* \\ \text{c)} & M_n(j\varphi)^T = M_n(\varphi)^* & \text{d)} & M_n(j\varphi j)^T = M_n(\varphi)^*. \end{array} \tag{4.6}$$

It is easily seen that for $\varphi = \begin{bmatrix}\alpha & \gamma\\ \beta & \delta\end{bmatrix}$ the latter conditions are equivalent to the following ones

$$\begin{array}{llll} \text{a)} & \alpha,\beta,\gamma,\delta \text{ real} & \text{b)} & \alpha = \overline{\gamma},\ \beta = \overline{\delta} \\ \text{c)} & \alpha = \overline{\beta},\ \gamma = \overline{\delta} & \text{d)} & \alpha = \overline{\delta},\ \beta = \overline{\gamma}\ , \end{array} \tag{4.7}$$

respectively. Let us denote the subclasses of $\mathcal{M}_n$ satisfying (4.7) by a) $\mathcal{M}_n(\mathbb{R},\mathbb{R})$ b) $\mathcal{M}_n(\mathbb{R},\mathbb{T})$ c) $\mathcal{M}_n(\mathbb{T},\mathbb{R})$ d) $\mathcal{M}_n(\mathbb{T},\mathbb{T})$, respectively.

The reason of this designation is the following circumstance. If $\varphi \in GL(\mathbb{C}^2)$ satisfies (4.7) then the corresponding Möbius transformation $\varphi(\lambda) = \frac{\alpha\lambda + \beta}{\gamma\lambda + \delta}$ acting within the Riemann sphere transforms a) $\mathbb{R}$ into $\mathbb{R}$, b) $\mathbb{R}$ into $\mathbb{T}$, c) $\mathbb{T}$ into $\mathbb{R}$, d) $\mathbb{T}$ into $\mathbb{T}$, respectively. Vice versa, if the function $\varphi(\lambda)$ is acting in such way then the matrix φ is of the form $\varphi = \theta\varphi^o$, where $\theta \in \mathbb{C}$ and φ^o belongs to the corresponding subclass of Möbius matrices.
Let us still mention that the classes $\mathcal{M}_n(\mathbb{R}, \mathbb{R})$ and $\mathcal{M}_n(\mathbb{T},\mathbb{T})$ are, obviously, subgroups of $\mathcal{M}_n$.
We obtained the following solution of Problem 2.

THEOREM 4.2. Let $y = Cx$ be a transformation of coordinates in $\mathbb{C}^n$. Then a) H-forms go over into H-forms, b) H-forms into T-forms, c) T-forms into H-forms, d) T-forms into T-forms if and only if a) $C \in \mathcal{M}_n(\mathbb{R}, \mathbb{R})$, b) $C \in \mathcal{M}_n(\mathbb{R},\mathbb{T})$, c) $C \in \mathcal{M}_n(\mathbb{T}, \mathbb{R})$, d) $C \in \mathcal{M}_n(\mathbb{T},\mathbb{T})$, respectively.

4.4. Transformations of Bezoutians and related classes of matrices.

The problems we stated at the beginning of this section are also relevant for other classes of matrices: Bezoutians (see 2.3), resultant matrices (see 2.5), difference and multiplication operators (see 0.3). Let us begin with the class of classical H-Bezoutians $\mathcal{B}^o(n)$.

THEOREM 4.3. A regular transformation $A \otimes B$ let the class $\mathcal{B}^o(n)$ invariant if and only if there is a matrix $\varphi \in GL(\mathbb{C}^2)$ such that $A = M_n(\varphi)$ and $B = M_n(\varphi)^T$. Moreover, the following equality holds

$$\mathrm{Bez}(M_{n+1}(\varphi)a, M_{n+1}(\varphi)b) = (\det \varphi) M_n(\varphi)\, \mathrm{Bez}(a,b)\, M_n(\varphi)^T. \tag{4.8}$$

PROOF. The sufficiency part of the theorem is proved if we show that (4.8) is fulfilled for the special Möbius matrices (4.4a), (4.4b). Suppose first $\varphi = a(\beta)$. Denoting by c_{ik} the entries of the matrix on the right side of (4.8) we have

$$c_{ik} = \sum_{r,s} \binom{r}{i}\beta^{r-i} b_{rs} \binom{s}{k}\beta^{s-k},$$

where b_{rs} are the entries of $\mathrm{Bez}(a,b)$. Therefore, we obtain for the generating function $C(\lambda,\mu)$ of the matrix $[c_{ik}]$

$$C(\lambda,\mu) = \sum_{r,s} b_{rs}(\lambda+\beta)^r(\mu+\beta)^s = B(\lambda+\beta,\mu+\beta),$$

where $B(\lambda,\mu)$ is the generating function of $\mathrm{Bez}(a,b)$. This implies (4.8).
If $\varphi = c(\alpha)$ then

$$C(\lambda,\mu) = \alpha B(\alpha\lambda,\alpha\mu),$$

hence (4.8) holds. In the case that $\varphi = j$ we have

$$C(\lambda,\mu) = - B(\lambda^{-1},\mu^{-1})\lambda^{n-1}\mu^{n-1},$$

which is just the generating function of $Bez(\hat{a},\hat{b})$. This completes the proof of sufficiency.

Assume now that $A \otimes B$ is a regular transformation invariant on $\mathcal{B}^o(n)$. That means, $A\ Bez(a,b)\ B$ is a Bezoutian for all $a(\lambda),\ b(\lambda) \in \mathbb{C}^{n+1}(\lambda)$. In case $Bez(a,b)$ is regular the matrix $B^{-1}\ Bez(a,b)^{-1}A^{-1}$ is Hankel. Since any regular H-matrix is the inverse of a Bezoutian we obtain that $B^{-1} \otimes A^{-1}$ transforms regular H-matrices into H-matrices. Since the class of regular $n\times n$ H-matrices is dense in $\mathcal{H}(n)$ $B^{-1} \otimes A^{-1}$ transforms $\mathcal{H}(n)$ into itself. Applying Theorem 4.1 we obtain that A^{-1} and also A is a Möbius matrix and $B = A^T$. The proof is complete. ≡

<u>COROLLARY 4.1.</u> A regular transformation $A \otimes B$ transforms the class of classical T-Bezoutians into the class of classical H-Bezoutians if and only if there is a $\varphi \in GL(\mathbb{C}^2)$ such that $A = M_n(\varphi)$ and $B = M_n(\varphi j)^T$. Moreover, the following equality holds

$$Bez(M_{n+1}(\varphi)a,M_{n+1}(\varphi)b) = (\det(\varphi j))M_n(\varphi)\ Bez_T(a,b)M_n(\varphi j)^T. \quad (4.8')$$

This equality immediately follows from (4.8) and the obvious relation

$$Bez_T(a,b)J_n = - Bez(a,b).$$

We leave to the reader the formulation of further corollaries and of a Bezoutian analogue of Theorem 4.2.

Let us now deal with multiplication operators $D_p(a)^T$, where $a(\lambda) \in \mathbb{C}^{n+1}(\lambda)$.

<u>PROPOSITION 4.2.</u> Suppose $\varphi \in GL(\mathbb{C}^2)$ and $a(\lambda) \in \mathbb{C}^{n+1}(\lambda)$, then

$$M_{n+p}(\varphi)D_p(a)^T M_p(\varphi^{-1}) = D_p(M_{n+1}(\varphi)a)^T. \quad (4.9)$$

<u>PROOF.</u> Denoting by M the left side of (4.9) we have for $x \in \mathbb{C}^p$

$$\begin{aligned} Mx(\lambda) &= M_{n+p}(\varphi)a(\lambda)x(\varphi^{-1}(\lambda))\left(\frac{\alpha - \gamma\lambda}{\det \varphi}\right)^{p-1} \\ &= a(\varphi(\lambda))x(\lambda)(\gamma\lambda + \delta)^n \\ &= (M_{n+1}(\varphi)a)(\lambda)x(\lambda), \end{aligned}$$

which yields (4.9). ≡

Basing on the equalities (4.8) and (4.9) we obtain a transformation rule for general Bezoutians.

<u>PROPOSITION 4.3.</u> Suppose $\varphi \in GL(\mathbb{C}^2)$, $a_1(\lambda),\ b_1(\lambda) \in \mathbb{C}^{m+1}(\lambda)$, $a_2(\lambda),\ b_2(\lambda) \in \mathbb{C}^{n+1}(\lambda)$ and denote $\tilde{a}_1(\lambda) := M_{m+1}(\varphi)a_1(\lambda)$, $\tilde{b}_1(\lambda) := M_{m+1}(\varphi)b_1(\lambda)$, $\tilde{a}_2(\lambda) := M_{n+1}(\varphi)a_2(\lambda)$, $\tilde{b}_2(\lambda) := M_{n+1}(\varphi)b_2(\lambda)$, then

$$Bez(\tilde{a}_1,\tilde{a}_2,\tilde{b}_1,\tilde{b}_2) = \det \varphi\ M_m(\varphi)\ Bez(a_1,a_2,b_1,b_2)\ M_n(\varphi)^T. \quad (4.10)$$

PROOF. According to Prop. 2.6 $B := \mathrm{Bez}(a_1,a_2,b_1,b_2)$ can be represented in the form

$$B = D_r(p)^T \,\mathrm{Bez}(a,b)\, D_r(q),$$

where $a_1(\lambda) = a(\lambda)p(\lambda)$, $a_2(\lambda) = a(\lambda)q(\lambda)$, $b_1(\lambda) = b(\lambda)p(\lambda)$, $b_2(\lambda) = b(\lambda)q(\lambda)$. Applying (4.8), (4.9) and denoting by M the right side of (4.10) we obtain

$$M = D_r(\tilde{p})^T(\det\varphi)\, M_r(\varphi)\,\mathrm{Bez}(a,b)\, M_r(\varphi)^T D_r(\tilde{q})$$

$$= D_r(\tilde{p})^T\,\mathrm{Bez}(\tilde{a},\tilde{b})\, D_r(\tilde{q}),$$

where $\tilde{p} = M_s(\varphi)p$, $\tilde{q} = M_t(\varphi)q$, $\tilde{a} = M_{r+1}(\varphi)a$, $\tilde{b} = M_{r+1}(\varphi)b$, and $s = m-r$, $t = n-r$. Taking into account that, obviously, $M_{m+1}(\varphi)a_1(\lambda) = \tilde{p}(\lambda)\tilde{a}(\lambda)$ and the analogous equalities for a_2, b_1, b_2 hold one concludes (4.10). ≡

Let us still remark a transformation rule for resultant matrices.

PROPOSITION 4.4. Suppose $\varphi \in GL(\mathbb{C}^2)$, $a(\lambda) \in \mathbb{C}^{m+1}(\lambda)$, $b(\lambda) \in \mathbb{C}^{n+1}(\lambda)$, $\tilde{a} = M_{m+1}(\varphi)a$, $\tilde{b} = M_{n+1}(\varphi)b$, then

$$\mathrm{Res}^1(\tilde{a},\tilde{b}) = \begin{bmatrix} M_{n+1}(\varphi^{-1})^T & 0 \\ 0 & M_{m+1}(\varphi^{-1})^T \end{bmatrix} \mathrm{Res}^1(a,b) M_{m+n+1}(\varphi)^T.$$

4.5. Root localization in the unit disc. The transformations regarded above can be used to deduce propositions about the root localization problems relative to the unit circle from such problems relative to the real axis. In particular, we obtain the Schur-Cohn theorem.

THEOREM 4.4. Let $c(\lambda)$ be a polynomial of degree n and $\mathrm{in}_{\mathbb{T}}c = \{\pi_+, \pi_-, \pi_0\}$ its inertia relative to the unit circle. Further let ϱ be the number of pairs of roots λ_0, μ_0 for which $\bar{\lambda}_0 = \mu_0^{-1}$. Then

$$\pi_{\pm} = p_{\pm}(C) + \varrho, \qquad \pi_0 = p_0(C) - 2\varrho, \tag{4.11}$$

where $(p_+(C), p_-(C), p_0(C))$ is the inertia of the Hermitian matrix

$$C = -\,\mathrm{Bez}_{\mathbb{T}}(c(\lambda), \hat{c}(\lambda)).$$

PROOF. For the proof we choose a Möbius matrix $M_n(\varphi) \in \mathcal{M}_n(\mathbb{R},\mathbb{T})$, where $\varphi = \begin{bmatrix}\alpha & \gamma \\ \beta & \delta\end{bmatrix}$. According to (4.7) we have $\gamma = \bar{\alpha}$, $\delta = \bar{\beta}$ and, therefore, $\det\varphi = 2i\,\mathrm{Im}\,\alpha\bar{\beta}$. We choose φ so that

$$\mathrm{Im}\,\alpha\bar{\beta} > 0.$$

In this case the function $\varphi(\lambda)$ transforms the upper half-plane into the unit disc. Therefore, we have

$$\mathrm{in}_{\mathbb{T}}(c) = \mathrm{in}_{\mathbb{R}}(M_{n+1}(\varphi)c). \tag{4.12}$$

Moreover, there is obviously

$$\varrho_{\mathbb{T}}(c) = \varrho_{\mathbb{R}}(M_{n+1}(\varphi)c), \tag{4.13}$$

where $\varrho_{\mathbb{T}}(\cdot)$ denotes the number of pairs of roots symmetric relative to $\mathbb{T}$ and $\varrho_{\mathbb{R}}(\cdot)$ the number of conjugate complex pairs of roots. In view of (4.8') we have

$$\mathrm{Bez}(M_{n+1}(\varphi)c,\ M_{n+1}(\varphi)\hat{c}) = 2i\ \mathrm{Im}\ \alpha\ \bar{\beta}\ M_n(\varphi)CM_n(\varphi)^*. \tag{4.14}$$

Furthermore, because of $M_{n+1}(\varphi j) = \overline{M_{n+1}(\varphi)}$ we have

$$M_{n+1}(\varphi)\hat{c} = \overline{M_{n+1}(\varphi)c}.$$

Setting $a := M_{n+1}(\varphi)c$ we obtain from (4.14)

$$\frac{1}{i}\ \mathrm{Bez}(a,\bar{a}) = 2\ \mathrm{Im}\ \alpha\bar{\beta}\ M_n(\varphi)CM_n(\varphi)^*.$$

It remains to apply Hermite's theorem (Theorem 2.4) and Sylvester's inertia law (Lemma 2.1), and the theorem is proved. ≡

Analogously to the case of upper half-plane the Schur-Cohn theorem solves the root localization problem for the unit disc only in case there are no roots of $c(\lambda)$ symmetric with respect to the unit circle. If there are such roots of $c(\lambda)$ then one can proceed in the same manner as in 2.7, 2.8. The first step will be the evaluation of the g.c.d. $g(\lambda)$ of $c(\lambda)$ and $\hat{c}(\lambda)$. This polynomial has, obviously, the property $\hat{g}(\lambda) = g(\lambda)$. Polynomials satisfying this condition will be called <u>symmetric.</u> It remains to solve now the root localization problem for the symmetric polynomial $g(\lambda)$.

<u>4.6. Root localization of symmetric polynomials.</u> The first idea for solving this problem is to use transformations of Bezoutians and to apply Theorem 2.5. Suppose $g(\lambda)$ is a symmetric polynomial of degree n and $\varphi = \begin{bmatrix}\alpha & \gamma\\ \bar{\alpha} & \bar{\gamma}\end{bmatrix}$. Then $M_n(\varphi) \in \mathcal{M}_n(T, R)$ and in view of (4.8')

$$\mathrm{Bez}_T(g,\tilde{g}) = -\,2i\ \mathrm{Im}\ \alpha\bar{\gamma}\ M_n(\varphi)\ \mathrm{Bez}(a,a')\ M_n(\varphi)^*, \tag{4.15}$$

where $a = M_{n+1}(\varphi^{-1})g$, $\tilde{g} = M_{n+1}(\varphi)a'$. Now we have to evaluate $\tilde{g}(\lambda)$.

<u>LEMMA 4.1.</u> Suppose $\varphi = \begin{bmatrix}\alpha & \gamma\\ \beta & \delta\end{bmatrix} \in GL(\mathbb{C}^2)$ and $a(\lambda) \in \mathbb{C}^{n+1}(\lambda)$ and denote $M := M_{n+1}(\varphi)$, then

$$(Ma')(\lambda) = \frac{1}{\det\varphi}[(\gamma\lambda+\delta)^2(Ma)'(\lambda) - n\gamma(\gamma\lambda+\delta)(Ma)(\lambda)]. \tag{4.16}$$

<u>PROOF.</u> We have

$$\begin{aligned}(Ma)'(\lambda) &= a'(\varphi(\lambda))(\gamma\lambda+\delta)^{n-2}\det\varphi + n\gamma(\gamma\lambda+\delta)^{n-1}a(\varphi(\lambda))\\ &= (\gamma\lambda+\delta)^{-2}\det\varphi(Ma')(\lambda) + n\gamma(\gamma\lambda+\delta)^{-1}(Ma)(\lambda).\end{aligned}$$

This yields (4.16). ≡

We may choose $\gamma = 1$, $\alpha = -i$ and obtain the following result.

<u>THEOREM 4.5.</u> Let $g(\lambda)$ be a symmetric polynomial and

$$g^{\tau}(\lambda) := i[(\lambda + 1)^2 g'(\lambda) - n(\lambda + 1)g(\lambda)].$$

Then

$$\mathrm{In}_{\mathbb{T}}(-i\ \mathrm{Bez}(g,g^{\tau})) = (\pi_0' + \varrho', \varrho', \delta), \tag{4.17}$$

where π_0' denotes the number of different roots of $g(\lambda)$ on $\mathbb{T}$ and ϱ' the number of different pairs of symmetric roots.

Let us mention another approach for solving the root localization problem basing on Lander's representation of the Bezoutian. We assume that the polynomial $g(\lambda) \in \mathbb{C}^{n+1}(\lambda)$ has simple non-zero roots $\lambda_1,\dots,\lambda_n$. Then we have for any $h(\lambda) \in \mathbb{C}^{n+1}(\lambda)$ according to Prop. 2.15

$$V_n(\lambda)^T\ \mathrm{Bez}_T(g,h)\ V_n(\lambda^{-1}) = \mathrm{diag}(\gamma_1,\dots,\gamma_n)\,, \tag{4.18}$$

where $\lambda := (\lambda_i)_1^n$, $\lambda^{-1} = (\lambda_i^{-1})_1^n$ and $\gamma_k = -\lambda_k^{-n+1}\, g'(\lambda_k)h(\lambda_k)$ $(k = 1,\dots,n)$.

We assume now that $g(\lambda)$ and $h(\lambda)$ are symmetric and the numbers λ_k are ordered as follows:

$$\lambda_1,\dots,\lambda_\pi \in \mathbb{T}\ ;\ \lambda_{\pi+2} = \bar{\lambda}_{\pi+1}^{-1},\dots,\lambda_n = \bar{\lambda}_{n-1}^{-1}.$$

Then, as it is easily seen, the matrix $C := -i\ \mathrm{Bez}_T(g,h)$ is Hermitian and

$$V_n^*(\lambda^{-1})\ C\ V_n(\lambda^{-1}) = \mathrm{diag}(\tfrac{\gamma_1}{i},\dots,\tfrac{\gamma_\pi}{i}, \begin{bmatrix} 0 & * \\ * & 0\end{bmatrix},\dots,\begin{bmatrix} 0 & * \\ * & 0\end{bmatrix}), \tag{4.19}$$

where the stars on the right side denote elements which are not important for the following. In particular, the numbers $-i\gamma_k$ $(k=1,\dots,\pi)$ are real. In order to solve the root localization problem by means of (4.19) we have to find a symmetric polynomial $h(\lambda)$ such that $-i\gamma_k$ $(k=1,\dots,\pi)$ are always positive. We shall show that
$h(\lambda) := i(\lambda g'(\lambda) - \frac{n}{2}\, g(\lambda))$ is such a polynomial. It is easily checked that $h(\lambda)$ is symmetric. To obtain that $-i\gamma_k > 0$ we introduce the following function on $[0,2\pi)$

$$H(\theta) = g(e^{i\theta})\ e^{-\frac{in}{2}\theta}.$$

Since $g(\lambda)$ is symmetric $H(\theta)$ will be a real-valued function. Furthermore, we have

$$H'(\theta) = h(e^{i\theta})\ e^{-\frac{in}{2}\theta}.$$

If $g(\lambda_k) = 0$ there are, moreover,

$$i\,\lambda_k\, g'(\lambda_k) = h(\lambda_k) \quad (k = 1,\dots,\pi)\,,$$

and we obtain

$$-i\gamma_k = \lambda_k^{-n} h(\lambda_k)^2 = H'(\theta_k)^2 > 0.$$

In this way we have proved the following result.

THEOREM 4.6. Let $g(\lambda)$ be a symmetric polynomial and

$$g^c(\lambda) = i(\lambda g'(\lambda) - \frac{n}{2} g(\lambda)); \quad C = -i \operatorname{Bez}_T(g, g^c).$$

then (4.19) holds.

We proved this theorem assuming that $g(\lambda)$ has simple roots only. If it has multiple roots one has to consider the polynomial $g(\lambda) + \varepsilon g^c(\lambda)$ for small ε, which has simple roots only and to repeat the proof.

5. KERNEL STRUCTURE

5.1. U-chains. A main tool for further investigations of T- and H-matrices is the fact that the kernel of such matrices has a special structure. To formulate these structure properties we need the concept of a U-chain, which is originated in the theory of Wiener-Hopf operators.

DEFINITION 5.1. A sequence of vectors $x_0, \dots, x_{r-1} \in \mathbb{C}^n$ is called *U-chain of length r* if there is a vector $p \in \mathbb{C}^s$ $(p \neq 0,\ s+r = n+1)$ such that

$$x_k = I_{ns}^k p \quad (k = 0, \dots, r-1).^{1)} \tag{5.1}$$

We shall say $x_0, \dots, x_{r-1}$ is a *U-chain generated by p*.

Obviously, the sequence $\{x_k\}_0^{r-1}$ is a U-chain if and only if

$$x_{k+1} = U_n x_k,$$

and the last components of x_k vanish for $k = 0, \dots, r-2$. We remark, furthermore, that a U-chain is always a linear independent system.

In the following very often occur subspaces, which are the linear hull of a U-chain. Such subspaces can also be characterized in the language of polynomials.

DEFINITION 5.2. Suppose

$$\Sigma = \{(\lambda_1, \nu_1), \dots, (\lambda_l, \nu_l)\}, \tag{5.2}$$

where $\lambda_i \in \overline{\mathbb{C}}$ and $\nu_i \in \mathbb{N}$, then $E^n(\Sigma)$ will denote the class of all $x \in \mathbb{C}^n$, for which the polynomial $x(\lambda)$ has roots λ_i of multiplicity $\geq \nu_i$. Here ∞ is said to be a root of multiplicity ν of $x(\lambda)$,

[1)] The matrices I_{ns}^k are defined in 0.2. For short we abbreviate $I^k = I_{ns}^k$.

$x = \{x_i\}_0^{n-1}$, if $x_{n-1} = \dots = x_{n-\nu} = 0$ and $x_{n-\nu-1} \neq 0$.
Clearly, $E^n(\Sigma)$ is a subspace of $\mathbb{C}^n$ for all systems Σ. It is nontrivial if $\sum \nu_i < n-1$.

PROPOSITION 5.1. A subspace E of $\mathbb{C}^n$ can be represented as the linear hull of a U-chain if and only if it can be represented in the form $E = E^n(\Sigma)$ for a system (5.2) with $\sum \nu_i < n-1$.

PROOF. Suppose $E \subseteq \mathbb{C}^n$ is the linear hull of the U-chain $I^0 p, \dots, I^{r-1} p$ $(p \in \mathbb{C}^s)$, then for the corresponding subspace $E(\lambda)$ of $C^n(\lambda)$ we have

$$E(\lambda) = \{c(\lambda)p(\lambda) : c(\lambda) \in \mathbb{C}^{r-1}(\lambda)\}.$$

Assuming

$$p(\lambda) = \gamma(\lambda-\lambda_1)^{\nu_1} \dots (\lambda-\lambda_{l-1})^{\nu_{l-1}}, \quad \nu_l = s-1-\sum_{i=1}^{l-1} \nu_i, \ \lambda_l = \infty \tag{5.3}$$

and setting $\Sigma = \{(\lambda_1, \nu_1), \dots, (\lambda_l, \nu_l)\}$ we obtain $E = E^n(\Sigma)$.
On the other hand, if $E = E^n(\Sigma)$, where Σ is of the form (5.2), $\lambda_l = \infty$ and $\sum \nu_i < n-1$, then the vector p corresponding to the polynomial $p(\lambda)$ defined by (5.3) generates a U-chain, which is just a basis of E. This completes the proof. ≡

5.2. The kernel structure theorem. In this subsection we shall characterize the kernel structure of T- and H-matrices in terms of U-chains. Clearly, it is sufficient to deal with one of these two classes. We choose the class of H-matrices.
Suppose $H = H_{mn}(b)$, where $b = (b_k)_0^{m+n-1}$. Besides the matrix H we consider the whole family of H-matrices generated by the vector b

$$H^k := H_{jk}(b) \quad (k = 1, \dots, m+n, \quad j+k = m+n). \tag{5.4}$$

We remark that

$$(H^k)^T = H^{m+n-k}. \tag{5.5}$$

The following obvious proposition shows the relevance of the matrices H^k for our investigation.

PROPOSITION 5.2. The U-chain $I^0 p, \dots, I^{r-1} p$ belongs to the kernel of H^k if and only if $p \in \ker H^{k-r+1}$.

We are going now to investigate the behaviour of the defect numbers of H^k. Denote

$$\alpha_k := \dim \ker H^k, \quad \beta_k := \dim \ker (H^k)^T.$$

Then we have, obviously,

$$\alpha_k \le \alpha_{k+1} \le \alpha_k + 2, \quad \beta_k - 2 \le \beta_{k+1} \le \beta_k \tag{5.6}$$

and

$$\alpha_k - \beta_k = 2k - m - n. \tag{5.7}$$

Suppose $\alpha_k > 0$ then we obtain according to Prop. 5.2 $I^0 p$, $I^1 p \in \ker H^{k+1}$ for any $p \in \ker H^k$. Hence $\alpha_{k+1} > \alpha_k$ and, in virtue of (5.7), $\beta_{k+1} > \beta_k - 2$. Analogously, $\beta_{k+1} > 0$ implies $\beta_{k+1} < \beta_k$ and $\alpha_{k+1} < \alpha_k + 2$. From these arguments we conclude the following assertion.

<u>PROPOSITION 5.3.</u> If $\alpha_{k+1} > 0$ and $\beta_k > 0$ then

$$\alpha_{k+1} = \alpha_k + 1, \quad \beta_{k+1} = \beta_k - 1. \tag{5.8a}$$

In case $\alpha_{k+1} = 0$ there is

$$\alpha_{k+1} = \alpha_k = 0, \quad \beta_{k+1} = \beta_k - 2, \tag{5.8b}$$

in case $\beta_k = 0$

$$\alpha_{k+1} = \alpha_k + 2, \quad \beta_{k+1} = \beta_k = 0. \tag{5.8c}$$

A successive application of Prop. 5.3 gives us the following result.

<u>PROPOSITION 5.4.</u> There are two numbers n_1 and n_2 such that

$$\alpha_{k+1} = 0, \quad \beta_{k+1} = \beta_k - 2 \quad \text{if} \quad 1 \leq k < n_1 \tag{5.9a}$$

$$\alpha_{k+1} = \alpha_k + 1, \quad \beta_{k+1} = \beta_k - 1 \quad \text{if} \quad n_1 \leq k < n_2 \tag{5.9b}$$

$$\alpha_{k+1} = \alpha_k + 2, \quad \beta_{k+1} = 0 \quad \text{if} \quad n_2 \leq k < m+n-1. \tag{5.9c}$$

Let us illustrate these facts in a picture.

<u>PICTURE 5.1.</u>

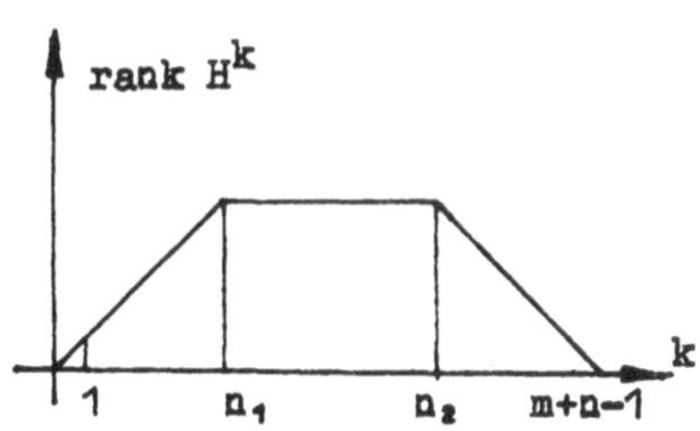

This implies

<u>COROLLARY 5.1.</u> In case $m+n$ is even we set $l := \frac{m+n}{2}$ and obtain

$$\operatorname{rank} H_{mn}(b) = \min\{m, n, \operatorname{rank} H_l(b)\}.$$

Otherwise, if $m+n$ is odd we set $l := \frac{m+n-1}{2}$ and obtain

$$\operatorname{rank} H_{mn}(b) = \min\{m, n, \operatorname{rank} H_{l,l+1}(b)\}.$$

<u>DEFINITION 5.3.</u> The numbers n_1 and n_2 will be called <u>characteristic degrees</u> of the matrix H or of the vector b.

Let us remark that in the trivial case $b = 0$ we have to put $n_1 = 0$, $n_2 = m+n$.

<u>COROLLARY 5.2.</u> For the numbers α_k $(k=1,\dots,n+m-1)$ hold

$$\alpha_k = \max\{k - n_1, 0\} + \max\{k - n_2, 0\}. \tag{5.10}$$

<u>THEOREM 5.1.</u> Let $H := H_{mn}(b)$ be an H-matrix with non-trivial kernel. Then the following assertions hold:

1. If H has no maximal rank, then the kernel of H and its transposed H^T is the linear hull of one U-chain generated by the same vector.
2. In general, the kernel of H can be represented as the linear hull of one or two U-chains.

<u>PROOF.</u> Consider the family of H-matrices H^k defined by (5.4). Let n_1 and n_2 be the characteristic degrees of H. In case H has no maximal rank we have, in view of Prop. 5.4, $n_1 < n < n_2$, in particular $n_1 \neq n_2$. Since $\alpha_{n_1+1} = 1$ we obtain

$$\ker H^{n_1+1} = \mathrm{lin}\{p\}$$

for some $p \in \mathbb{C}^{n_1+1}$. By Prop. 5.2, this implies

$$\ker H \supseteq \mathrm{lin}\{I^0 p, \dots, I^{n-n_1-1} p\}.$$

Respecting that U-chains are linearly independent systems and $\dim \ker H = n - n_1$ we observe that equality holds in the latter inclusion. Taking into account still (5.5) we conclude the first assertion of the theorem.

Suppose now H has maximal rank. Then $n \geq n_2$. Obviously, in case $n = n_2$ the kernel of H is the linear hull of one U-chain, analogously to the case of non-maximal rank. Consider the case $n > n_2$. We have here $I^0 p, \dots, I^{n_2-n_1} p \in \ker H^{n_2+1}$ and $\alpha_{n_2+1} = n_2 - n_1 + 2$. Hence there is a vector $p \in \mathbb{C}^{n_2+1}$ such that

$$\ker H^{n_2+1} = \mathrm{lin}\{I^0 p, \dots, I^{n_2-n_1} p, q\}.$$

This implies, by Prop. 5.2, for $k = n_2+1, \dots, m+n-1$

$$\ker H^k \supseteq \mathrm{lin}\{I^0 p, \dots, I^{k-n_1-1} p,\ I^0 q, \dots, I^{k-n_2-1} q\}. \tag{5.11}$$

In virtue of Cor. 5.2 equality holds in the inclusion (5.11) if the vectors on the right side are linearly independent. This can be proved by induction. We assume that the system of the two U-chains on the right side of (5.11) is linearly independent for k but linearly dependent for $k+1$. Then there are numbers $\alpha, \beta, \alpha_i, \beta_i$ such that $|\alpha| + |\beta| > 0$ and

$$\alpha I^{k-n_1}p + \beta I^{k-n_2}q = \sum_{i=0}^{k-n_1-1} \alpha_i I_p^i + \sum_{i=0}^{k-n_2-1} \beta_i I_q^i .$$

We put $\tilde{q} := \alpha I^{n_2-n_1}p + \beta I^o q$, then $I^o\tilde{q},\dots,I^{k-n_2-1}\tilde{q} \in \ker H^k$. Since the right side of the latter equality obviously belongs to $\ker H^k$ we have, moreover,

$$I^{k-n_2}\tilde{q} = \alpha I^{k-n_1}p + \beta I^{k-n_2}q \in \ker H^k.^{1)}$$

According to Prop. 5.2 this implies $\tilde{q} \in \ker H^{n_2}$, which is a contradiction to our assumption.
We have proved that equality holds in (5.11). Choosing $k = n$ we obtain the second assertion, and the proof of the theorem is complete. ≡

<u>REMARK 5.1.</u> With the same arguments as in the proof of Theorem 5.1 one can show that the system of vectors

$$I^o p,\dots,I^{n_2-1}p, \quad I^o q,\dots,I^{n_1-1}q$$

is linearly independent. Since the matrix with these columns is just the transposed resultant matrix of $p(\lambda)$ and $q(\lambda)$ this is equivalent to the regularity of $\mathrm{Res}(p,q)$.

<u>COROLLARY 5.3.</u> The kernel of a non-regular square H- or T-matrix can be represented as the linear hull of a U-chain.

Let us formulate this corollary also in the language of polynomials: The kernel of a H- or T-matrix is always a subspace of the form $E^n(\Sigma)$.

<u>5.3. Characteristic polynomials and inverse problems.</u> In the previous subsection we defined the concept of characteristic degrees n_1 and n_2 of an H-matrix. In case $n_1 < n_2$ we have a vector p such that

$$\ker H^{n_1+1} = \mathrm{lin}\{p\}$$

and a vector q such that

$$\ker H^{n_2+1} = \mathrm{lin}\{I^o p,\dots,I^{n_2-n_1}p,q\},$$

provided that $n_2 < m+n-1$. If $n_2 = m+n-1$ (or, which is the same, $n_1 = 1$) we take for q an arbitrary vector of $\mathbb{C}^{m+n}$ linearly independent of $I^o p,\dots,I^{m+n-2}p$. In case $n_1 = n_2$ there are, by definition, two vectors p, q such that

$$\ker H^{n_1+1} = \mathrm{lin}\{p,q\}.$$

<u>DEFINITION 5.4.</u> The two polynomials $p(\lambda)$, $q(\lambda)$ corresponding to the vectors p, q defined above will be called <u>characteristic polynomials of the H-matrix $H_{mn}(b)$</u> or <u>of the vector b</u>.

---

1) In these considerations a vector $(x_k)_o^{n-1}$ is identified with all vectors $(x_k)_o^{n-1+p}$, where $x_n = \dots = x_{n-1+p} = 0$.

Let us discuss the uniqueness of characteristic polynomials. In case $n_1 < n_2$ the polynomial $p(\lambda)$ is unique up to a constant factor, $q(\lambda)$ is unique up to a linear combination of $p(\lambda),\dots,\lambda^{n_2-n_1} p(\lambda)$. In case $n_1 = n_2$ any basis of $\ker H^{n_1+1}$ corresponds to a pair of characteristic polynomials.

We remark that the degrees of the characteristic polynomials are not equal to the characteristic degrees, in general. More precisely, if the degree of $p(\lambda)$ is equal to n_1 then $q(\lambda)$ may be chosen such that its degree is n_1-1.

If the degree of $p(\lambda)$ is smaller than n_1 then n_2 coincides with the degree of $q(\lambda)$ for all $q(\lambda)$ (see Prop. 5.4).

The characteristic polynomials of a regular H-matrix have closed relations to the fundamental solutions. Indeed, let y and w be the solutions of the fundamental equations (1.20), then $p(\lambda) = y(\lambda)$ and $q(\lambda) = w(\lambda) - \lambda^n$ are characteristic polynomials and $n_1 = n_2 = n$.

Another pair of characteristic polynomials is $y(\lambda)$ and $\lambda y(\lambda)-y^{n-1}(\lambda)$, where $H_{n-1}y^{n-1} = e_{n-2}$, provided that the matrix $H_{n-1}(b)$ is regular.

Analogously, for regular T-matrices $T_n(a)$ a pair of characteristic polynomials is $x(\lambda)$ and $z(\lambda) - \lambda^n$. Moreover, provided that $T_{n-1}(a)$ is regular another pair is given by $x(\lambda)$ and $y(\lambda)$. Here $x(\lambda)$, $z(\lambda)$, $y(\lambda)$ denote the polynomials corresponding to the fundamental solutions of (1.3), (1.4), (1.5), respectively.

An immediate consequence of Remark 5.1 and Cor. 2.5 is the following assertion.

<u>PROPOSITION 5.5.</u> Let $p(\lambda) \in \mathbb{C}^{n_1+1}(\lambda)$ and $q(\lambda) \in \mathbb{C}^{n_2+1}(\lambda)$ be characteristic polynomials of an H-matrix. Then

$$\sigma(p) \cap \sigma(q) = \emptyset. \text{ [1]} \tag{5.12}$$

We shall show that the converse assertion is also true.

<u>THEOREM 5.2.</u> Suppose $p(\lambda) \in \mathbb{C}^{n_1+1}(\lambda)$ and $q(\lambda) \in \mathbb{C}^{n_2+1}(\lambda)$ are polynomials satisfying (5.12). Then there exists a vector $b \in \mathbb{C}^{n_1+n_2-1}$ such that $p(\lambda)$ and $q(\lambda)$ are characteristic polynomials of b (i.e. of all $H_{mn}(b)$, $m+n = n_1+n_2$).

The proof of the theorem bases on the following fact.

<u>LEMMA 5.1.</u> The polynomials $p(\lambda)$, $q(\lambda)$ are characteristic ones of b if and only if

$$\mathrm{Res}^{-1}(p,q)b = 0 \tag{5.13}$$

and (5.12) hold.

<u>PROOF.</u> Suppose $p(\lambda)$ and $q(\lambda)$ are characteristic ones of b, then, by Prop. 5.2, the U-chains

$$I^0 p,\dots,I^{n_2-2}p,\ I^0 q,\dots,I^{n_1-2}q$$

[1] Recall that $\sigma(a)$ denotes the set of roots of $a(\lambda)$ <u>inclusively</u> $\underline{\infty}$.

belong to the kernel of $H_{1,n_1+n_2-1}(b) = b^T$. This is equivalent to (5.13). Furthermore, by Prop. 5.5, the relation (5.12) holds. Consequently, the necessity part is proved.

Let us now assume that both (5.12) and (5.13) are satisfied. We have to show that $p(\lambda)$ and $q(\lambda)$ are characteristic. Suppose, for a moment, they are not. In this case the characteristic degrees n_1^o and n_2^o fulfil inequalities $n_1^o < n_1$ and, in virtue of $n_1 + n_2 = n_1^o + n_2^o$, $n_2^o > n_2$. Let $p_o(\lambda)$, $q_o(\lambda)$ be a pair of characteristic polynomials of b. Then p and q can be represented as a linear combination of the U-chain generated by p_o, since $n_1, n_2 < n_2^o$. In other words this means, there are polynomials $s(\lambda)$ and $t(\lambda)$ such that $p(\lambda) = s(\lambda)p_o(\lambda)$ and $q(\lambda) = t(\lambda)p_o(\lambda)$. Hence $p_o(\lambda)$ is a common divisor of $p(\lambda)$ and $q(\lambda)$. This contradicts (5.12). The contradiction shows that $p(\lambda)$ and $q(\lambda)$ are characteristic, and the lemma is proved. ≡

For the proof of Theorem 5.2 we have merely to choose a vector $b \in \ker \mathrm{Res}^{-1}(p,q)$ and to apply Lemma 5.1. ≡

<u>REMARK 5.2.</u> Let $p(\lambda)$, $q(\lambda)$ be polynomials satisfying (5.12). Then the vector b, for which $p(\lambda)$, $q(\lambda)$ are characteristic is uniquely determined up to a multiplicative constant.

Indeed, in case $p(\lambda)$ and $q(\lambda)$ are characteristic polynomials the rows of $\mathrm{Res}^{-1}(p,q)$ are, as shown above, linearly independent. This implies

$$\dim \ker \mathrm{Res}^{-1}(p,q) = 1,$$

and the remark is proved.

From Theorem 5.2 and Prop. 5.5 we deduce some results about inverse problems for the fundamental equations. Here, by the term "inverse problem" is meant a problem of the following form: For two given vectors we ask if there is an H-matrix (or a T-matrix) such that these vectors are solutions of the fundamental equations. Of course, there are various variants of inverse problems.

<u>THEOREM 5.3.</u> For given vectors $y,w \in \mathbb{C}^n$ there exists a vector $b = (b_k)_0^{2n-1}$ such that

$$H_n(b)y = e_{n-1}, \quad H_n(b)w = h, \tag{5.14}$$

where $h = (b_{n+k})_0^{n-1}$, if and only if

$$\sigma(y(\lambda)) \cap \sigma(w(\lambda) - \lambda^n) = \emptyset. \tag{5.15}$$

In case the latter condition is satisfied the vector b is uniquely determined, and $H_n(b)$ is regular.

<u>PROOF.</u> Suppose that there exists a vector b such that y,w are fundamental solutions. Then the polynomials $y(\lambda)$ and $w(\lambda) - \lambda^n$ are characteristic and, therefore, by Prop. 5.5, (5.15) is satisfied.

Assume now that (5.15) is fulfilled. Then, according to Theorem 5.2, there is a vector $b' = (b'_k)_0^{2n-2}$ such that $y(\lambda)$ and $w(\lambda) - \lambda^n$ are characteristic polynomials of b'. This implies

$$H_n(b')y = \gamma e_{n-1}, \quad H_n(b')w = h',$$

where $h' = (b'_{n+k})_0^{n-1}$ and b'_{2n-1} is some number uniquely determined by b'. We have $\gamma \neq 0$, since otherwise the vector $\tilde{b}' := \begin{bmatrix} b' \\ b_{2n-1} \end{bmatrix}$ belongs to the kernel of $\mathrm{Res}(y, w(\lambda) - \lambda^n)$, which contradicts (5.15). Putting now $b := \frac{1}{\gamma} \tilde{b}'$ (5.14) is fulfilled. Since b' is, by Remark 5.2, uniquely determined up to a constant factor we obtain that b is uniquely determined, too. The regularity of $H_n(b)$ follows from Theorem 1.1', and thus the theorem is completely proved. ≡

<u>THEOREM 5.4.</u> For given vectors $y = (y_k)_0^{n-1} \in \mathbb{C}^n$ and $y' \in \mathbb{C}^{n-1}$, where $y_{n-1} \neq 0$, there exists a vector $b \in \mathbb{C}^{2n-1}$ such that

$$H_n(b)y = e_{n-1}, \quad H_{n-1}(b)y' = e_{n-2} \tag{5.16}$$

if and only if

$$\sigma(y(\lambda)) \cap \sigma(y'(\lambda)) = \emptyset. \tag{5.17}$$

In case the latter condition is satisfied b is uniquely determined, and the matrices $H_n(b)$ and $H_{n-1}(b)$ are regular.

<u>PROOF.</u> Suppose that $b \in \mathbb{C}^{2n-1}$ is such that (5.16) is fulfilled. Then, in virtue of Theorem 1.4' and $y_{n-1} \neq 0$, $H_n(b)$ and $H_{n-1}(b)$ are regular[1)]. Therefore, $y(\lambda)$, $\lambda y(\lambda) - y'(\lambda)$ is a pair of characteristic polynomials of b. According to Prop. 5.5 we obtain

$$\sigma(y(\lambda)) \cap \sigma(\lambda y(\lambda) - y'(\lambda)) = \emptyset, \tag{5.17'}$$

which is equivalent to (5.17).

Assume now that (5.17) is fulfilled. Then (5.17') holds, too. By Theorem 5.2 there is a vector $b' \in \mathbb{C}^{2n-1}$ such that $y(\lambda)$ and $\lambda y(\lambda) - y'(\lambda)$ are characteristic polynomials of b'. We obtain

$$H_n(b')y = \alpha e_{n-1}, \quad H_{n-1}(b')y' = \alpha e_{n-2}.$$

Since $\alpha = 0$ yields $(b_k)_0^{2n-3} \in \ker \mathrm{Res}(y, y')$, which contradicts (5.17), we have $\alpha \neq 0$. Therefore, the vector $b := \frac{1}{\alpha} b'$ fulfils the equations (5.16). The uniqueness of b follows from Remark 5.2, the regularity of $H_n(b)$ and $H_{n-1}(b)$ from Theorem 1.4'. The proof is complete. ≡

The next variant of inverse problems we shall formulate for T-matrices.

<u>THEOREM 5.5.</u> For given vectors $x = (x_k)_0^{n-1}$, $y = (y_k)_0^{n-1} \in \mathbb{C}^n$, where $x_0 \neq 0$, there exists a vector $a = (a_k)_{1-n}^{n-1}$ such that

---

1) Obviously, the solution v of Theorem 1.4' is given by $v(\lambda) = y'(\lambda) - \beta y(\lambda)$, where $\beta = [b_{n-1} \cdots b_1]y'$.

$$T_n(a)x = e_o, \quad T_n(a)y = e_{n-1} \tag{5.18}$$

if and only if

$$\sigma(x(\lambda)) \cap \sigma(y(\lambda)) = \emptyset \tag{5.19}$$

and

$$x_o = y_{n-1}. \tag{5.20}$$

If these conditions are satisfied then the vector a is uniquely determined and $T_n(a)$, $T_{n-1}(a)$ are regular.

PROOF. If the vector $a \in \mathbb{C}^{2n-1}$ is such that (5.18) is satisfied then $x(\lambda)$ and $\lambda y(\lambda)$ are characteristic polynomials of a. Therefore, (5.19) holds. The equality (5.20) we remarked in Subsection 1.2.
Suppose now that (5.19) and (5.20) are fulfilled. Then there is a vector $a' \in C^{2n-1}$ such that $x(\lambda)$ and $\lambda y(\lambda)$ are characteristic polynomials of a'. This implies

$$T_n(a')x = \alpha e_o, \quad T_n(a')y = \beta e_{n-1}.$$

We have $\alpha \neq 0$ and $\beta \neq 0$ since otherwise $(a'_k)_{2-n}^{n-1} \in \ker \operatorname{Res}(x,y)$ or $(a'_k)_{1-n}^{n-2} \in \ker \operatorname{Res}(x,y)$, which contradicts (5.19). Moreover, it is easily seen that $\frac{1}{\alpha} x_o = \frac{1}{\beta} y_{n-1}$. Hence, according to (5.20) we have $\alpha = \beta$. It remains to take $a := \frac{1}{\alpha} a'$, and the theorem is proved. ≡

5.4. Partial indices. In the theory of infinite block Toeplitz matrices (i.e. matrix-valued Wiener-Hopf operators) the concept of partial indices plays an important role. This concept can be also introduced for finite T-matrices and H-matrices. It is remarkable that, different to the infinite case, already <u>scalar</u> finite T- or H-matrices have <u>two</u> partial indices.
In order to give a definition of this concept we recall our results about the kernel structure of an H-matrix H:
If H has no maximal rank then

$$\ker H = \operatorname{lin}\{I^o p,\dots,I^{\varkappa_1-1}p\}; \ \ker H^T = \operatorname{lin}\{I^o p,\dots,I^{\nu_2-1}p\}.$$

In case H has maximal rank we have

$$\ker H = \operatorname{lin}\{I^o p,\dots,I^{\varkappa_1-1}p,\ I^o q,\dots,I^{\varkappa_2-1}q\}, \ \ker H^T = \{0\}$$

or

$$\ker H = \{0\}, \ \ker H^T = \operatorname{lin}\{I^o p,\dots,I^{\nu_2-1}p, I^o q,\dots,I^{\nu_1-1}q\}.$$

Here $p(\lambda)$, $q(\lambda)$ denote a pair of characteristic polynomials. Clearly, the numbers $\varkappa_1$, $\varkappa_2$, ν_1, ν_2 are uniquely determined. We set $\varkappa_2 := -\nu_2$ in the first and $\varkappa_i := -\nu_i$ $(i = 1,2)$ in the third case.

DEFINITION 5.5. To any H-matrix H we associate the pair of integers $\varkappa(H) = (\varkappa_1, \varkappa_2)$. The numbers $\varkappa_1$, $\varkappa_2$ will be called partial indices of H.[1) In particular, H is regular if and only if $\varkappa(H) = (0,0)$.

By definition we have always $\varkappa_2 \leq \varkappa_1$. Let us remark some further elementary properties.

PROPOSITION 5.6. Suppose $H \in \mathcal{H}(m,n)$, then:

1) $\varkappa_1 + \varkappa_2 = n - m$

2) $\dim \ker H = \max\{\varkappa_1, 0\} + \max(\varkappa_2, 0\}$,

$\dim \ker H^T = \max\{-\varkappa_1, 0\} + \max\{-\varkappa_2, 0\}$

3) $\varkappa_1 = n - n_1$, $\varkappa_2 = n - n_2$.

Consider now for a given vector $b \in \mathbb{C}^{N-1}$ the family of H-matrices $H^k = H_{jk}(b)$ $(j+k = N)$.

PROPOSITION 5.7.

$$\varkappa(H^{k+1}) = \varkappa(H^k) + (1,1). \tag{5.21}$$

The proof of this proposition is obtained by applying Prop. 5.2, Prop. 5.3 and Remark 5.1.

5.5. Solvability of the fundamental equations. In this subsection we show that there is a closed connection between the solvability of fundamental equations of an H-matrix and the kernel structure of its transposed. Supposing $H := H_{mn}(b)$ is an H-matrix, $b = (b_k)_0^{m+n-2}$ we consider the following called fundamental equations

$$Hw(\alpha) = h(\alpha) \qquad (\alpha \in \overline{\mathbb{C}}), \tag{5.22}$$

where $h(\alpha) := (b_{k+n})_0^{m-1}$, $b_{m+n-1} = \alpha$ if $\alpha \in \mathbb{C}$ and $h(\infty) = e_{m-1}$. We remark that the linear hull of $h(\alpha)$ is the two-dimensional subspace of $\mathbb{C}^m$ spanned by e_{m-1} and $h(0)$.

THEOREM 5.6. Let $\varkappa_1$, $\varkappa_2$ denote the partial indices of H.

1. If (5.22) is solvable for two different $\alpha = \alpha_1$ and $\alpha = \alpha_2$ then $\varkappa_1, \varkappa_2 \geq 0$ (i.e. $\ker H^T = \{0\}$). In the latter case (5.22) is solvable for all α.
2. The equation (5.22) is solvable for exactly one α if and only if $\varkappa_1 \geq 0 > \varkappa_2$.
3. There is not any $\alpha \in \overline{\mathbb{C}}$ such that the equation (5.22) has a solution if and only if $0 > \varkappa_1 \geq \varkappa_2$.

The proof bases on the following fact well-known from the linear algebra:

---

1) A quite general definition of this concept is given in Part II.

The equation $Ax = y$ is solvable if and only if $u^T y = 0$ for all $u \in \ker A^T$.

In our case (5.22) is solvable if and only if $u^T h(\alpha) = 0$ for all $u \in \ker H^T$.

We split the proof of Theorem 5.6 into some parts.

<u>LEMMA 5.2.</u> If $\varkappa_1 \geq 0 > \varkappa_2$ then there exists exactly one α such that (5.22) is solvable.

<u>PROOF.</u> By assumption we have $\ker H^T \neq \{0\}$ and $\ker H^T$ is spanned by a U-chain

$$I^o p, \dots, I^l p.$$

It is easily seen that $I^k p$ $(k = 0,\dots,l-1)$ fulfil the condition $(I^k p)^T h(\alpha) = 0$ for all α. Therefore, (5.22) is solvable if and only if the linear functional $f := (I^l p)^T$ vanishes on $h(\alpha)$. Since the vectors $h(\alpha)$ span a two-dimensional subspace there is at least one α such that $f(h(\alpha)) = 0$. It remains to show that f is not identically zero on the space spanned by the vectors $h(\alpha)$. Suppose $f(h(\alpha)) = 0$ for all α. Then, in particular, $f(h(\infty)) = 0$, which means that the last component of $I^l p$ vanishes. This together with $f(h(o)) = 0$ implies, as it is easily seen, $I^{l+1} p \in \ker H^T$, which contradicts our assumption. Hence, $f(h(\alpha))$ vanishes for exactly one α. ≡

<u>LEMMA 5.3.</u> If $0 > \varkappa_1 \geq \varkappa_2$ then (5.22) is insolvable for all $\alpha \in \overline{\mathbb{C}}$.

<u>PROOF.</u> By assumption $\ker H^T$ has a basis of two U-chains

$$I^o p, \dots, I^l p, \quad I^o q, \dots, I^j q, \tag{5.23}$$

where $p(\lambda)$ and $q(\lambda)$ are characteristic polynomials of H. Since, by Prop. 5.5, $\sigma(p) \cap \sigma(q) = \emptyset$ the highest order coefficients of $p(\lambda)$ and $q(\lambda)$ do not vanish simultaneously. Hence, without restriction of generality we may assume that the highest order coefficient of one polynomial, say $q(\lambda)$, is zero and of the other, $p(\lambda)$, is non-zero. It is easily verified that

$$(I^k p)^T h(\alpha) = (I^i q)^T h(\alpha) = 0$$

for $k = 0,\dots,l-1$; $i = 0,\dots,j-1$ and all α. Therefore (5.22) is solvable if and only if the linear functionals $f := (I^l p)^T$ and $g := (I^j q)^T$ vanish on $h(\alpha)$. Since, by assumption, the last component of $I^j q$ is equal to zero we conclude from $g(h(\alpha)) = 0$ that $I^{j+1} q \in \ker H^T$. Since the vector $I^{j+1} q$, by Remark 5.1, is linearly independent of the system (5.23) this is a contradiction to our assumption. Hence $g(h(\alpha)) \neq 0$, which means that (5.22) is not solvable, and the proof is complete. ≡

<u>PROOF OF THEOREM 5.6.</u>

1. Suppose (5.22) is solvable for $\alpha = \alpha_1$ and $\alpha = \alpha_2$ $(\alpha_1 \neq \alpha_2)$.

Then it is solvable for all α, since any $h(\alpha)$ is a linear combination of $h(\alpha_1)$ and $h(\alpha_2)$. We have to show $\ker H^T = \{0\}$. Let u be an element of $\ker H^T$. Then $u^T h(\infty) = 0$, which means the last component of u vanishes. Moreover, $u^T h(0) = 0$ implies that the shifted vector $U_m u$ also belongs to $\ker H^T$. Repeating these arguments we obtain $u = 0$, and the first part of the theorem is proved.

2. If (5.22) is solvable for exactly one α then $\ker H^T \neq \{0\}$, which is $0 > \varkappa_2$. We have furthermore $\varkappa_1 \geq 0$, since $\varkappa_1 < 0$ implies, by Lemma 5.3, that (5.22) is insolvable. The converse direction we obtain by Lemma 5.2.

3. The third part of Theorem 5.6 follows from Lemma 5.3 and the second part. ≡

It is remarkable that the solutions of the fundamental equations can be expressed with the help of characteristic polynomials.

PROPOSITION 5.8.

1. Suppose $\varkappa_1 \geq \varkappa_2 > 0$ and let $p(\lambda)$, $q(\lambda)$ be a pair of characteristic polynomials of $H := H_{mn}(b)$. Then for any α there are $\lambda(\alpha)$, $\mu(\alpha)$ such that

$$w(\alpha) = \lambda(\alpha)\, I^{\varkappa_1} p + \mu(\alpha)\, I^{\varkappa_2} q + u, \qquad u \in \ker H$$

is the general solution of (5.22).

2. Suppose $\varkappa_1 > 0 \geq \varkappa_2$ and let $p(\lambda)$ be the characteristic polynomial with the lowest characteristic degree and the highest order coefficient p_{n_1}. If $p_{n_1} = 0$ then (5.22) is solvable for $\alpha = \infty$ (and only for this α). If $p_{n_1} \neq 0$ then (5.22) is solvable for

$$\alpha = -\frac{1}{p_{n_1}}[b_{m-1},\dots,b_{m+n-2}]\, I^{\varkappa_1} p.$$

In both cases the general solution of (5.22) for this α is represented by

$$w(\alpha) = -\frac{1}{p_{n_1}}\, I^{\varkappa_1} p + u, \qquad u \in \ker H.$$

The proof of this proposition is obtained by a straightforward calculation.

We recall that $u \in \ker H$ in Prop. 5.8, 1. means

$$u = \sum_{k=0}^{\varkappa_1 - 1} \lambda_k\, I_p^k + \sum_{j=0}^{\varkappa_2 - 1} \mu_j\, I^j q$$

and in Prop. 5.8 , 2.

$$u = \sum_{k=0}^{\varkappa_1 - 1} \lambda_k\, I^k p, \qquad \text{where } \lambda_k, \mu_j \text{ are arbitrary.}$$

REMARK 5.3. The numbers $\lambda(\alpha)$ and $\mu(\alpha)$ occurring in Prop. 5.8 can be expressed by means of p and q, too.

We still mention a connection between the partial indices and some modified fundamental equations. Denote by G_k the linear hull of the vectors $(U_m^T)^i h(\alpha)$ for $i = 0,\ldots,k-1$; $\alpha \in \overline{\mathbb{C}}$ (clearly, it is sufficient to take $\alpha = 0,\infty$). Consider the "modified" fundamental equations

$$Hw_\varkappa(\alpha) = (U_m^T)^\varkappa h(\alpha) + h_\varkappa \quad (\varkappa = 1,2,\ldots), \tag{5.24a}$$

where $h_\varkappa \in G_\varkappa$ is arbitrary. Obviously, if (5.24a) has a solution for some $k = \varkappa$ then it has a solution for all $k < \varkappa$. Beside this "weaker" fundamental equations we consider "stronger" fundamental equations

$$\begin{cases} Hw_\varkappa(\alpha) = h(\alpha) \\ w_0 = \ldots = w_{-\varkappa-1} = 0 \end{cases} \quad (\varkappa = 0,-1,\ldots). \tag{5.24b}$$

Furthermore, it is not difficult to verify the following assertion.

PROPOSITION 5.9. Suppose $\varkappa_1, \varkappa_2$ are the partial indices of H. Then (5.24) is solvable for all α if $\varkappa \geq -\varkappa_2$, solvable for exactly one α if $-\varkappa_2 > \varkappa \geq -\varkappa_1$ and insolvable for all α if $-\varkappa_1 > \varkappa$.

5.6. (ϱ,π)-characteristic of H-matrices and (ϱ,π_+,π_-)-characteristic of T-matrices. In the next subsection we shall study the kernel behaviour after extending of an H- or of a T-matrix. For this it is useful to define the concepts of (ϱ,π)-characteristic of an H-matrix and (ϱ,π_+,π_-)-characteristic of a T-matrix first introduced (but in another form) by I. S. Iochvidov. Let us note that there are essential differences between H- and T-matrices concerning such extension problems.

Suppose $p(\lambda)$ is a characteristic polynomial of $H := H_{mn}(b)$ corresponding to the lowest characteristic degree n_1. We denote by π the multiplicity of the root ∞ of $p(\lambda)$, respectively, we put $\pi := 0$ if ∞ is no root of $p(\lambda)$. In other words, π is the difference between the characteristic degree n_1 and the degree of $p(\lambda)$ in the usual sense. Further we define

$$\varrho := \max\{r : H_r(b) \text{ is regular}\}.$$

DEFINITION 5.6. The pair of natural numbers $\operatorname{char} H := (\varrho,\pi)$ will be called (ϱ,π)-characteristic of H.

Let us remark that the pair (ϱ,π) merely depends on the vector b, not on the type of the matrix H. Furthermore, it is clear that regular n×n H-matrices have the characteristic $(n,0)$. Therefore, we obtain also $\pi = 0$ if the partial indices of H coincide (see Prop. 5.7).

For example if $b := (\lambda^i)_0^{m+n-2}$ we have $\operatorname{char} H_{mn}(b) = (1,0)$.
Now let $p(\lambda)$ be a characteristic polynomial corresponding to the lowest characteristic degree of a T-matrix $T := T_{mn}(a)$. Then we denote by π_- the multiplicity of the root ∞ and by π_+ the multiplicity of the root 0.

<u>DEFINITION 5.7.</u> The triple $\operatorname{char} T := (\varrho, \pi_+, \pi_-)$ will be called <u>(ϱ,π_+,π_-)-characteristic of T</u>.
Analogously to H-matrices a regular nxn T-matrix has the characteristic $(n,0,0)$.

<u>THEOREM 5.7.</u> Let (ϱ, π) be the characteristic of an H-matrix $H_{mn}(b)$. Then

$$\operatorname{rank} H_{mn}(b) = \min\{m, n, \pi+\varrho\}. \tag{5.25}$$

<u>PROOF.</u> According to Cor. 5.1 it is sufficient to prove the proposition in case $0 \leq n-m \leq 1$. Let us assume first $m = n$ and let $p(\lambda)$ be a characteristic polynomial corresponding to the lowest characteristic degree. If $H_n(b)$ is regular then (5.25) is trivial. Suppose $H_n(b)$ is singular. Then

$$\ker H_n(b) = \operatorname{lin}\{I^0 p, \ldots, I^{\varkappa-1} p\},$$

where $\operatorname{rank} H_n(b) = n - \varkappa$. We introduce an nxn matrix

$$P = [e_0 \ldots e_{\tau-1} \quad I^0 p \ldots I^{\varkappa-1} p \quad I^{\varkappa} p \ldots I^{\varkappa+\pi-1} p], \tag{5.26}$$

where $\tau = n - \varkappa - \pi$. Obviously, P is regular and an upper triangular matrix. Furthermore, it is easy to verify that

$$P^T H_n(b) P = \begin{bmatrix} H_\tau(b) & 0 \\ 0 & \Gamma \end{bmatrix}, \tag{5.27}$$

where Γ is an H-matrix of the form

$$\Gamma = \begin{bmatrix} 0 & 0 \\ 0 & \Gamma_0 \end{bmatrix} \begin{matrix} \}\varkappa \\ \}\pi \end{matrix}, \quad \Gamma_0 = \begin{bmatrix} 0 & & \gamma_0 \\ & \cdot^{\cdot^{\cdot}} & \vdots \\ \gamma_0 & \cdots & \gamma_{\pi-1} \end{bmatrix} \quad (\gamma_0 \neq 0).$$

Denoting by C_k the principal section $[c_{ij}]_0^{k-1}$ of a matrix $C = [c_{ij}]_0^n$ and by $\tilde{H}$ the matrix on the right side of (5.27) we obtain from (5.27)

$$P_k^T H_k(b) P_k = \tilde{H}_k \quad (k = 1, \ldots, n). \tag{5.27'}$$

The matrices $\tilde{H}_k$ are singular for $k > \tau$. On the other side, since $\dim \ker \tilde{H} = \varkappa$, the matrix $H_\tau(b)$ must be regular. We conclude $\varrho = \tau$, which implies (5.25).
In case $n = m+1$ (5.25) is proved in a similar way if one considers in (5.27) the matrix $P_{n-1}^T H_{n-1,n}(b) P$. This completes the proof. ≡

THEOREM 5.8. Let (ϱ,π_+,π_-) be the characteristic of a T-matrix $T_{mn}(b)$. Then

$$\operatorname{rank} T_{mn}(a) = \min\{m,n,\pi_+ + \pi_- + \varrho\}. \tag{5.28}$$

PROOF. We prove (5.28) for singular square T-matrices. The case $m = n-1$ can be verified analogously. For general m, n (5.28) results in view of Cor. 5.1. Let $p(\lambda)$ be a characteristic polynomial corresponding to the lowest characteristic degree of $T_n(a)$. Then

$$\ker T_n(a) = \operatorname{lin}\{I^0p,\dots,I^{\varkappa-1}p\}$$

and, in view of Prop. 0.2,

$$\ker T_n(a)^T = \operatorname{lin}\{J_nI^0p,\dots,J_nI^{\varkappa-1}p\}.$$

We introduce the following matrices

$$P = [e_0\dots e_{\tau-1}\ \ I^0p\dots I^{n-\tau-1}p],$$

$$Q = [e_0\dots e_{\tau-1}\ \ J_nI^{n-\tau-1}p\dots J_nI^0p],$$

where $\tau = n - \varkappa - \pi_+ - \pi_-$. Obviously, P, Q are upper triangular matrices and

$$Q^TT_n(a)P = \begin{bmatrix} T_\tau(a) & 0 \\ 0 & \Gamma \end{bmatrix},$$

where Γ is a T-matrix of the form

$$\Gamma = \begin{bmatrix} 0 & 0 & \Gamma_- \\ 0 & 0 & 0 \\ \Gamma_+ & 0 & 0 \end{bmatrix} \begin{matrix} \}\pi_- \\ \}\varkappa \\ \}\pi_+ \end{matrix}, \quad \Gamma_- = \begin{bmatrix} \gamma_0^- & \cdots & \gamma_{\pi_--1}^- \\ & \ddots & \vdots \\ 0 & & \gamma_0^- \end{bmatrix}, \quad \Gamma_+ = \begin{bmatrix} \gamma_0^+ & & 0 \\ \vdots & \ddots & \\ \gamma_{\pi_+-1}^+ & \cdots & \gamma_0^+ \end{bmatrix}$$

$$(\gamma_0^\pm \neq 0).$$

Analogously to the proof of Theorem 5.7 we obtain $\varrho = \tau$, which implies (5.28). The theorem is completely proved. ≡

For our further consideration it will be convenient to define the concept of quasi-regularity.

DEFINITION 5.8. An H-matrix will be called <u>quasi-regular</u> if $\pi = 0$. A T-matrix will be called <u>quasi-regular</u> if $\pi_+ = \pi_- = 0$. It will be called <u>(+)-half-regular</u> (<u>(-)-half-regular</u>) if $\pi_+ = 0$ $(\pi_- = 0)$. Clearly, regular T- and H-matrices are also quasi-regular. Furthermore, if an H-matrix $H_{mn}(b)$ is quasi-regular then this is also valid for all H-matrices $H^{n+k} = H_{m-k,n+k}(b)$ $(k = -n+1,\dots,m-1)$. An analogous statement holds for T-matrices. From this we conclude that T- or H-matrices for which the partial indices coincide are quasi-regular. In case they do not coincide we have in the family of matrices H^{n+k} at least one matrix for which the partial indices $\varkappa_1,\varkappa_2$ possess the property $\varkappa_1 > 0 \geq \varkappa_2$. For these matrices we have the following obvious criterion of quasi-regularity.

PROPOSITION 5.10. Suppose for the partial indices $\varkappa_1$, $\varkappa_2$ of the H-matrix H $\varkappa_1 > 0 \geq \varkappa_2$. Then H is quasi-regular if and only if there is a vector in the kernel of H with non-vanishing last component. If $\varkappa_1$, $\varkappa_2$ denotes the partial indices of a T-matrix T and $\varkappa_1 > 0 \geq \varkappa_2$ then T is (+)-half-regular ((-)-half-regular) if and only if there is a vector in ker T with non-vanishing first (last) component.

THEOREM 5.9. Any H-matrix $H \in \mathcal{H}(m,n)$ with char $H = (\varrho,\pi)$ admits a unique representation

$$H = H_0 + H_\infty, \tag{5.29}$$

where char $H_0 = (\varrho,0)$ and char $H_\infty = (0,\pi)$.

PROOF. As noted above it is sufficient to prove the theorem for the case $0 \leq n-m \leq 1$. Suppose $m = n$ and $H = H_n(b)$, $b = (b_k)_0^{2n-2}$. Further, let $p(\lambda) = p_0 + p_1\lambda + \ldots + p_\varrho\lambda^\varrho$ be a characteristic polynomial corresponding to the lowest characteristic degree. Obviously, we have

$$H[I^0p\ldots I^{n-\varrho-1}p] = \begin{bmatrix} 0 & 0 \\ \underbrace{0}_{\varkappa} & \Sigma \end{bmatrix}\}{n-\pi} \qquad (\varkappa = n-\varrho-\pi), \tag{5.30}$$

where Σ is an H-matrix of the form

$$\Sigma = \begin{bmatrix} 0 & & \sigma_0 \\ & \cdot^{\cdot^{\cdot}} & \vdots \\ \sigma_0 & \cdots & \sigma_{\pi-1} \end{bmatrix} \qquad (\sigma_0 \neq 0). \tag{5.31}$$

We put

$$\Pi := \begin{bmatrix} p_\varrho & \cdots & p_{\varrho-\pi+1} \\ & \ddots & \vdots \\ 0 & & p_\varrho \end{bmatrix} \tag{5.32}$$

and choose $H_\infty := H_n(w)$, where $w_0 = \ldots = w_{2n-2-\pi} = 0$, and the other w_k's are defined by

$$\begin{bmatrix} 0 & & w_{2n-\pi-1} \\ & \cdot^{\cdot^{\cdot}} & \vdots \\ w_{2n-\pi-1} & \cdots & w_{2n-2} \end{bmatrix} = \Sigma\,\Pi^{-1}.$$

We have now by definition

$$(H - H_\infty)[I^0p\ldots I^{n-\varrho-1}p] = 0 \tag{5.33}$$

and conclude from this that $H - H_\infty$ is quasi-regular. Consequently, the existence of a representation (5.29) is shown. The case $n-m = 1$ can be proved in an analogous way. It remains to prove the uniqueness.

Let $\tilde{H}_\infty$ be another matrix with char $\tilde{H}_\infty = (0,\pi)$ such that $\tilde{H}_0 := H - \tilde{H}_\infty$ is quasi-regular. Then $\tilde{H}_\infty$ has the form

$$\tilde{H}_\infty = \begin{bmatrix} 0 & 0 \\ 0 & W \end{bmatrix}, \tag{5.34}$$

and $\tilde{H}_0$ fulfils

$$\tilde{H}_0[I^0 p \ldots I^{n-\varrho-1}p] = 0.$$

We obtain $\Sigma - W\,\Pi = 0$, hence $\tilde{H}_\infty = H_\infty$, and the theorem is completely proved. ≡

Let us present the analogous theorem for T-matrices, which can be shown similarly.

THEOREM 5.10. Any T-matrix T with char $T = (\varrho,\pi_+,\pi_-)$ admits a unique representation

$$T = T_0 + T_+ + T_-, \tag{5.35}$$

where char $T_0 = (\varrho,0,0)$, char $T_+ = (0,\pi_+,0)$, char $T_- = (0,0,\pi_-)$.

5.7. Partial indices after one-row or one-column extensions.

For simplifying the forthcoming studies let us recollect some relations between the quantities introduced in the latter subsections.

TABLE 5.1.

Notations		
$H \in \mathcal{H}(m,n)$		$T \in \mathcal{T}(m,n)$
	n_1, n_2 characteristic degrees	$(n_1 \le n_2)$
	$\varkappa_1\ \varkappa_2$ partial indices	$(\varkappa_1 \ge \varkappa_2)$
char $H = (\varrho,\pi)$		char $T = (\varrho,\pi_+,\pi_-)$
$\alpha = \dim\ker H$		$\alpha = \dim\ker T$
$\beta = \dim\ker H^T$		$\beta = \dim\ker T^T$
	Relations [1]	
	$n_1 + n_2 = m + n$	
	$\varkappa_1 = n - n_1 = n_2 - m$, $\varkappa_2 = n - n_2 = n_1 - m$	
	$\varkappa_1 - \varkappa_2 = n_2 - n_1$, $\varkappa_1 + \varkappa_2 = n - m$	
	$\alpha = \max\{\varkappa_1,0\} + \max\{\varkappa_2,0\}$	
	$\beta = \max\{-\varkappa_1,0\} + \max\{-\varkappa_2,0\}$	
	$r = \min\{m,n,n_1\}$	
$n_1 = \pi + \varrho$		$n_1 = \pi_+ + \pi_- + \varrho$

[1] See especially Pict. 5.1, Cor. 5.1 and 5.2, Prop. 5.6, and Theorem 5.7.

Throughout this subsection $H = H_{mn}(b)$, $b = (b_i)_0^{m+n-2}$, denotes an H-matrix with characteristic polynomials $p(\lambda) \in \mathbb{C}^{n_1+1}(\lambda)$ and $q(\lambda) \in \mathbb{C}^{n_2+1}(\lambda)$. We consider the following one-row or one-column extensions of H

$$H_+(\alpha) := H_{m+1,n}(b(\alpha)), \quad H_-(\alpha) := H_{m,n+1}(b(\alpha)),$$

where $b(\alpha) := (b,\alpha) \in \mathbb{C}^{m+n}$, $\alpha \in \mathbb{C}$. Further, we denote by $n_1^+(\alpha)$, $n_2^+(\alpha)$ or $n_1^-(\alpha)$, $n_2^-(\alpha)$ the characteristic degrees and by $\varkappa_1^+(\alpha)$, $\varkappa_2^+(\alpha)$ or $\varkappa_1^-(\alpha)$, $\varkappa_2^-(\alpha)$ the partial indices of $H_+(\alpha)$ or $H_-(\alpha)$, respectively. We ask about relations between $\varkappa_i$ and $\varkappa_i^+(\alpha)$, $\varkappa_i^-(\alpha)$ $(i = 1,2)$.

THEOREM 5.11.
(1) Let H be quasi-regular and $n_1 < n_2$.[1] Then there exists exactly one value $\alpha = \alpha_0$ such that $H_+(\alpha_0)$ is quasi-regular again and

$$\varkappa_1^+(\alpha_0) = \varkappa_1 , \quad \varkappa_2^+(\alpha_0) = \varkappa_2 - 1, \tag{5.36a}$$

$$n_1^+(\alpha_0) = n_1 , \quad n_2^+(\alpha_0) = n_2 + 1. \tag{5.36b}$$

For $\alpha \neq \alpha_0$ $H_+(\alpha)$ is not quasi-regular and

$$\varkappa_1^+(\alpha) = \varkappa_1 - 1 , \quad \varkappa_2^+(\alpha) = \varkappa_2, \tag{5.37a}$$

$$n_1^+(\alpha) = n_1 + 1 , \quad n_2^+(\alpha) = n_2. \tag{5.37b}$$

(2) In case H is not quasi-regular and $n_1 < n_2$ there exists no α such that $H_+(\alpha)$ is quasi-regular and (5.37) holds.
(3) If $n_1 = n_2$ then $H_+(\alpha)$ is quasi-regular for all α and (5.36) holds.

PROOF. We consider the families of matrices

$$H^k := H_{m+n-k,k}(b) \quad \text{and} \quad H_\alpha^k := H_{m+n-k+1,k}(b(\alpha)).$$

First let us assume that $n_1 < n_2$. In this case we have

$$\ker H_\alpha^{n_1+1} \subseteq \ker H^{n_1+1} = \operatorname{lin}\{p\}$$

and $p \in \ker H_\alpha^{n_1+1}$ if and only if

$$p_{n_1}\alpha = - [b_{n_2-2} \cdots b_{m+n-2}]p', \tag{5.38}$$

where $p = (p'p_{n_1})$. If H is quasi-regular, then $p_{n_1} \neq 0$ and, therefore, (5.38) is satisfied if and only if

$$\alpha = \alpha_0 := - p_{n_1}^{-1}[b_{n_2-2} \cdots b_{m+n-2}]p'.$$

Hence we have $\varkappa_1^+(\alpha_0) = \varkappa_1$ and if $\alpha \neq \alpha_0$, $\varkappa_1^+(\alpha) = \varkappa_1 - 1$. In both cases $p(\lambda)$ is a characteristic polynomial, but in the second case

[1] We recall that $n_1 < n_2$ is equivalent to $\varkappa_1 > \varkappa_2$ (c.f. Table 5.1)

we must $p(\lambda)$ consider as an element of $\mathbb{C}^{n_1+2}(\lambda)$. That means $H_+(\alpha)$ is quasi-regular if and only if $\alpha = \alpha_0$. In virtue of Prop. 5.6 the first assertion is proved.

Now let H be not quasi-regular. Then $p_{n_1} = 0$ and (5.38) is not satisfied for all α, since otherwise $p' \in \ker H^{n_1+2}$ and, therefore, $I^0p', I^1p' \in \ker H^{n_1+1}$, which contradicts our assumption. In this way we proved (5.37).

It remains to consider the case $n_1 = n_2$. In this case we have

$$\ker H_\alpha^{n_1+1} \subseteq \ker H^{n_1+1} = \text{lin}\{p,q\}$$

and, moreover, H^{n_1} is regular. Therefore, $\dim \ker H_\alpha^{n_1+1} = 1$ for all α, where the kernel is spanned by a linear combination $\tilde{p}$ of p and q. The highest order coefficient of $\tilde{p}(\lambda)$ is non-zero, since otherwise we had a non-trivial kernel of H^{n_1}. Hence $H_\alpha^{n_1+1}$ is always quasi-regular, and the theorem is proved. ≡

From the proof of Theorem 5.11 it becomes clear that in case $n_1 < n_2$ the characteristic polynomial $p(\lambda)$ of H is also a characteristic polynomial of $H_{\pm}(\alpha)$. Furthermore, there are characteristic polynomials $p(\lambda), q(\lambda)$ of H which are characteristic polynomials of $H_{\pm}(\alpha)$, too. Therefore the following assertion is true.

COROLLARY 5.4.

(1) Suppose $\varkappa_1\varkappa_2 \leq 0$, then $\ker H_+(\alpha) = \ker H$ if and only if H is quasi-regular and $\alpha = \alpha_0$, but

$$\ker H_+(\alpha) = \text{lin}\{I^0p,\dots,I^{\varkappa_1-2}p\}$$

otherwise.

(2) In case $\varkappa_1 > \varkappa_2 > 0$ there exists a characteristic polynomial $q(\lambda)$ of H such that

$$\ker H = \text{lin}\{I^0p,\dots,I^{\varkappa_1-1}p,\ I^0q,\dots,I^{\varkappa_2-1}q\},$$

$$\ker H_+(\alpha) = \text{lin}\{I^0p,\dots,I^{\varkappa_1-1}p,\ I^0q,\dots,I^{\varkappa_2-2}q\}$$

if and only if H is quasi-regular and $\alpha = \alpha_0$, but

$$\ker H_+(\alpha) = \text{lin}\{I^0p,\dots,I^{\varkappa_1-2}p,\ I^0q,\dots,I^{\varkappa_2-1}q\}$$

otherwise.

From Cor. 5.4 one may conclude the following assertion about the behaviour of the (ϱ,π)-characteristic of H-matrices after extension.

COROLLARY 5.5. Suppose $\text{char } H = (\varrho,\pi)$, then

$$\text{char } H_+(\alpha) = \text{char } H_-(\alpha) = \begin{cases} (\varrho,\ 0) \text{ if } & \varkappa_1 = \varkappa_2 \text{ or } \pi = 0,\ \alpha = \alpha_0 \\ (\varrho,\pi+1) & \text{otherwise}. \end{cases}$$

REMARK 5.4. For H-matrices one-column extension and one-row extension are equivalent, because we have

$$H_-(\alpha) = H_\alpha^{n+1}.$$

Therefore, the following relations hold

$$\varkappa_i^+(\alpha) = \varkappa_i^-(\alpha) - 1, \quad n_i^+(\alpha) = n_i^-(\alpha) \quad (i = 1,2).$$

Remark 5.4 fails to be true for T-matrices.
We are going to show this fact. Let $T = T_{mn}(a)$, $a = (a_i)_{1-n}^{m-1}$, denote a T-matrix with characteristic polynomials $p(\lambda)$, $q(\lambda)$ corresponding to the characteristic degrees n_1, n_2.
We consider one-row or one-column extension of T

$$T_+(\alpha) := T_{m+1,n}(a_+(\alpha)), \quad T_-(\alpha) := T_{m,n+1}(a_-(\alpha)),$$

where $a_+(\alpha) = (a,\alpha)$ and $a_-(\alpha) = (\alpha,a)$. As above we shall mark the partial indices and characteristic degrees of the extensions with α.

THEOREM 5.12.
(1) Let T be (+)-half-regular ((−)-half-regular) and $n_1 < n_2$. Then there exists exactly one value $\alpha = \alpha_0$ such that $T_+(\alpha_0)$ is (+)-half-regular ($T_-(\alpha_0)$ is (−)-half-regular) and

$$\begin{gathered} \varkappa_1^+(\alpha_0) = \varkappa_1, \ \varkappa_2^+(\alpha_0) = \varkappa_2 - 1, \\ (\varkappa_1^-(\alpha_0) = \varkappa_1 + 1, \ \ \varkappa_2^-(\alpha_0) = \varkappa_2). \end{gathered} \tag{5.39}$$

For $\alpha \neq \alpha_0$ $T_+(\alpha)$ is not (+)-half-regular ((−)-half-regular) and

$$\varkappa_1^+(\alpha) = \varkappa_1 - 1, \ \varkappa_2^+(\alpha) = \varkappa_2, \quad (\varkappa_1^-(\alpha) = \varkappa_1, \ \ \varkappa_2^-(\alpha) = \varkappa_2 + 1) \ , \tag{5.40}$$

(2) If T is not (+)-half-regular ((−)-half-regular) and $n_1 < n_2$, then $T_+(\alpha)$ is not (+)-half-regular ($T_-(\alpha)$ not (−)-half-regular) and (5.40) holds.

(3) In case $n_1 = n_2$ $T_+(\alpha)$ is (+)-half-regular and $T_-(\alpha)$ (−)-half-regular for all α and (5.39) holds.

The proof is analogous to that one of Theorem 5.11. ≡

Let us formulate corollaries analogously to Cor. 5.4 and 5.5.

COROLLARY 5.4'.
(1) Suppose $\varkappa_1\varkappa_2 \leqslant 0$, then $\ker T = \ker T_+(\alpha)$ if and only if T is (+)-half-regular and $\alpha = \alpha_0$, but

$$\ker T_+(\alpha) = \mathrm{lin}\{I^0p,\ldots,I^{\varkappa_1-2}p)$$

otherwise.

(2) In case $\varkappa_1 > \varkappa_2 > 0$ there exists a characteristic polynomial $q(\lambda)$ of T such that

$$\ker T = \mathrm{lin}\{I^0p,\ldots,I^{\varkappa_1-1}p, \ I^0q,\ldots,I^{\varkappa_2-1}q\},$$

$$\ker T_+(\alpha) = \mathrm{lin}\{I^0p,\ldots,I^{\varkappa_1-1}p, \ I^0q,\ldots,I^{\varkappa_2-2}q\}$$

if and only if T is (+)-half-regular and $\alpha = \alpha_0$, but

$$\ker T_+(\alpha) = \mathrm{lin}\{I^0p,\ldots,I^{\varkappa_1-2}p, \ I^0q,\ldots,I^{\varkappa_2-1}q\}$$

otherwise.

COROLLARY 5.5'. Suppose $\operatorname{char} T = (\varrho,\pi_+,\pi_-)$, then

$$\operatorname{char} T_+(\alpha) = \begin{cases} (\varrho,0,\pi_-) & \text{if } \varkappa_1 = \varkappa_2 \text{ or } \pi_+ = 0,\ \alpha = \alpha_0 \\ (\varrho,\pi_++1,\pi_-) & \text{otherwise} \end{cases},$$

$$\operatorname{char} T_-(\alpha) = \begin{cases} (\varrho,\pi_+,0) & \text{if } \varkappa_1 = \varkappa_2 \text{ or } \pi_- = 0,\ \alpha = \alpha_0 \\ (\varrho,\pi_+,\pi_-+1) & \text{otherwise} \end{cases}.$$

5.8. Singular extensions. Let A be an $m\times n$ matrix. An $(m+\mu)\times(n+\nu)$ matrix $\mathfrak{X}$, which has A as a submatrix in its north-west corner will be called (μ,ν)-extension of A, i.e. if $\mathfrak{X} = [a_{ik}]_1^{m+\mu}\,{}_1^{n+\nu}$ then $A = [a_{ik}]_1^m\,{}_1^n$. Clearly, $\operatorname{rank} \mathfrak{X} \geq \operatorname{rank} A$.

DEFINITION 5.9. An extension $\mathfrak{X}$ of A with the property $\operatorname{rank} \mathfrak{X} = \operatorname{rank} A$ will be refered to as a singular extension of A.

We consider now the problem whether an H- or T-matrix has singular extensions, which are H- or T-matrices again. Speaking about extensions of H- and T-matrices we shall always mean such extensions.
By applying Theorem 5.11 and 5.12 successively we obtain the following results.

THEOREM 5.13. Let A be a quasi-regular H- or T-matrix with the partial indices $\varkappa_1$, $\varkappa_2$ satisfying the condition $\varkappa_1\varkappa_2 \leq 0$. Then for all $\mu,\nu = 0,1,2,\ldots$ there exists a singular (μ,ν)-extension of A. This extension is uniquely determined if and only if $\varkappa_1 \neq \varkappa_2$. On the other side, suppose A has a singular (μ,ν)-extension for some $\mu,\nu > 0$, then A is quasi-regular and $\varkappa_1\varkappa_2 \leq 0$.

REMARK 5.5. In case that A is Hankel the quasi-regularity of A follows already provided that there is a singular $(\mu,0)$- or $(0,\nu)$-extension $(\mu,\nu > 0)$ of A. This statement fails to be true for T-matrices.

In the Toeplitz case we obtain the following.

THEOREM 5.14. Let T be a (+)-half-regular ((−)-half-regular) T-matrix with the partial indices satisfying the condition $\varkappa_1\varkappa_2 < 0$. Then for all $\mu = 1,2,\ldots$ there exists a singular $(\mu,0)$-extension ($(0,\mu)$-extension) of T, which is unique if and only if $\varkappa_1 \neq \varkappa_2$. On the other side, suppose T has a singular $(\mu,0)$-extension ($(0,\mu)$-extension) for some $\mu > 0$, then T is (+)-half-regular ((−)-half-regular) and $\varkappa_1\varkappa_2 \leq 0$.

DEFINITION 5.10. An (μ,ν)-extension $T_{m+\mu,n+\nu}(a)$ of a T-matrix $T_{mn}(a)$ will be called (+)-half-singular ((−)-half-singular) if $T_{m+\mu,n}(a)$ ($T_{m,n+\nu}(a)$) is a singular extension of $T_{mn}(a)$.
We deduce now some further conclusions from Theorem 5.11 and 5.12. Let us consider (1,1)-extensions $H[1] := H_{m+1,n+1}(b)$, $T[1] := T_{m+1,n+1}(a)$

of an H- or T-matrix $H_{mn}(b)$, $T_{mn}(a)$, respectively. We study the behaviour of the kernel, the partial indices and the characteristic under the following assumption

$$\varkappa_1(k) \neq \varkappa_2(k) \quad (k = 0,\dots,l-1), \tag{5.41}$$

where $\varkappa_1(k)$, $\varkappa_2(k)$ denotes the partial indices of $H[k]$ (respectively $T[k]$). The condition (5.41) is fulfilled, in particular, if all matrices $H[k]$ (respectively $T[k]$) for $k = 0,\dots,l-1$ have no maximal rank.

PROPOSITION 5.11. Suppose $H[l]$ is an $(l\times l)$-extension of an H-matrix H and the condition (5.41) is satisfied. Then there is a number l_1 such that

$$\varkappa_1(k+1) = \varkappa_1(k) + 1, \quad \varkappa_2(k+1) = \varkappa_2(k) - 1 \quad \text{if } 0 \leq k < l_1,$$

$$\varkappa_1(k+1) = \varkappa_1(k) - 1, \quad \varkappa_2(k+1) = \varkappa_2(k) + 1 \quad \text{if } l_1 < k < l$$

holds. If $k = l_1$ then the latter relation holds or there is

$$\varkappa_1(l_1+1) = \varkappa_1(l_1), \quad \varkappa_2(l_1+1) = \varkappa_2(l_1).$$

Let us formulate the analogous proposition for T-matrices.

PROPOSITION 5.12. Suppose $T[l]$ is an $(l\times l)$-extension of a T-matrix T and (5.41) is satisfied. Then there are two numbers l_1, l_2 such that

$$\varkappa_1(k+1) = \varkappa_1(k) + 1, \quad \varkappa_2(k+1) = \varkappa_2(k) - 1 \quad \text{if } 0 \leq k < l_1,$$

$$\varkappa_1(k+1) = \varkappa_1(k), \quad \varkappa_2(k+1) = \varkappa_2(k) \quad \text{if } l_1 \leq k < l_2,$$

$$\varkappa_1(k+1) = \varkappa_1(k) - 1, \quad \varkappa_2(k+1) = \varkappa_2(k) + 1 \quad \text{if } l_2 \leq k < l.$$

These two propositions can be easily verified, and their proofs will be omitted.
Now, let us illustrate the typical behaviour of the partial indices (and therefore also of the kernel dimension and the rank) after arbitrary kxk-extension of an arbitrary H-matrix.

PICTURE 5.1.[1]

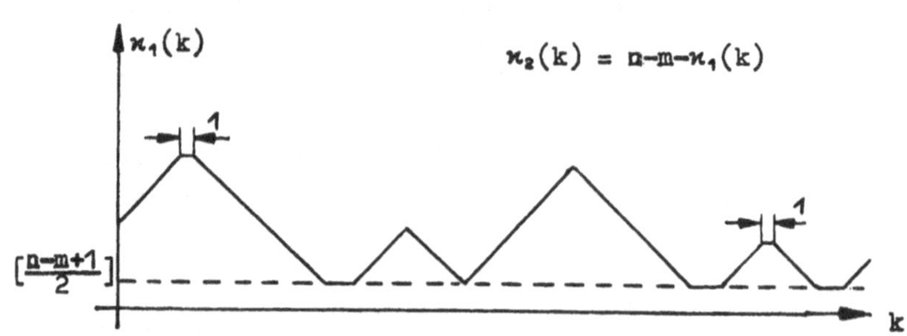

[1] $[\cdot]$ denotes the entire-function.

Different to this picture the typical behaviour of these quantities for T-matrices looks as follows.

PICTURE 5.2.

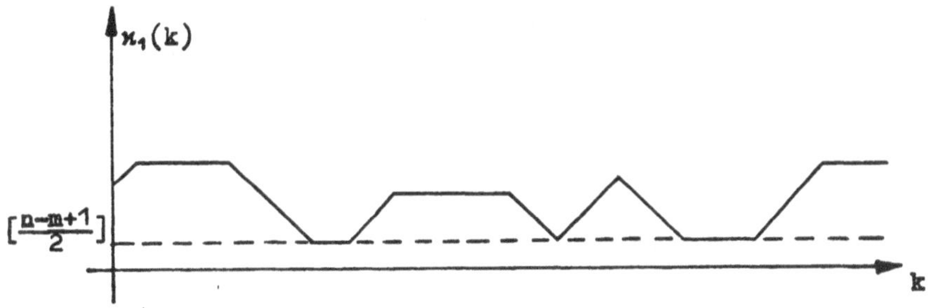

5.9. Kernel structure of square H- and T-matrices. In Section 6 we shall need the following kernel description of square H- and T-matrices.

THEOREM 5.15. Suppose $H = H_n(b)$ is a non-regular H-matrix, $\mathrm{char}\, H = (\varrho,\pi)$ and

$$\ker H_{\varrho+1}(b) = \mathrm{lin}\{u\}. \tag{5.42}$$

Then

$$\ker H = \mathrm{lin}\{I^0 u,\dots,I^{n-\pi-\varrho-1}u\}. \tag{5.43}$$

PROOF. We know that (c.f. Table 5.1) $r = \pi + \varrho$ is the lowest characteristic degree of H. That means

$$\ker H^{r+1} = \mathrm{lin}\{p\},$$

where $H^{r+1} := H_{2n-r-1,r+1}(b)$ and $p(\lambda)$ is a characteristic polynomial of H. By the definition of the (ϱ,π)-characteristic the last π components of p vanish, i.e. $p = (p',0)$, where $p' \in \mathbb{C}^{r+1}$. Since, obviously, $H_{\varrho+1}(b)p' = 0$, we obtain $p' = u$ (up to a multiplicative constant, which is insignificant). Therefore,

$$H_{2n-r-1,\varrho+1}(b)u = 0.$$

In virtue of Prop. 5.2, this is equivalent to $I^k u \in \ker H$ for $k = 0,\dots,n-r-1$. In that manner we proved that the inclusion "$\supseteq$" in (5.43) holds. The converse inclusion results if we take into account Theorem 5.7 and the linear independence of a U-chain. The theorem is proved. ≡

THEOREM 5.16. Suppose $T = T_n(a)$ is a non-regular T-matrix, $\mathrm{char}\, T = (\varrho,\pi_+,\pi_-)$ and

$$\ker T_{\varrho+1}(a) = \mathrm{lin}\{u\}.$$

Then

$$\ker T = \mathrm{lin}\{I^{\pi_+}u,\dots,I^{n-\varrho-\pi_--1}u\}.$$

The proof is analogous to that one of Theorem 5.15.
The converse statements of Theorem 5.15 and 5.16 are also valid. More precisely speaking the following assertions hold true.

PROPOSITION 5.13. Suppose $H = H_n(b)$ and

$$\ker H = \operatorname{lin}\{I^0 u, \dots, I^{\alpha-1} u\},$$

where $u \in \mathbb{C}^{\varrho+1}$, and the last component of u is non-zero. Then

$$\operatorname{char} H = (\varrho, n-\varrho-\alpha). \tag{5.44}$$

PROOF. By Prop. 5.2 we have

$$\ker H^{n-\alpha+1} = \operatorname{lin}\{u\}.$$

Therefore, $u(\lambda)$ is a characteristic polynomial of H with the characteristic degree $n - \alpha$. Since the last component of u is assumed to be non-zero we obtain $\pi = n - \varrho - \alpha$. Taking into account still $n_1 = \pi + \varrho$ (c.f. Table 5.1) we obtain (5.44). ≡

PROPOSITION 5.13'. Suppose $T = T_n(a)$ and

$$\ker T = \operatorname{lin}\{I^s u, \dots, I^{t-1} u\},$$

where $u \in \mathbb{C}^{\varrho+1}$ with non-vanishing first and last component. Then

$$\operatorname{char} T = (\varrho, s, n-t-\varrho).$$

5.10. Strongly singular T- and H-matrices. A last aim of this section is to give a first description of some classes of infinite T- and H-matrices.

DEFINITION 5.11. An infinite matrix $A = [a_{ik}]_0^\infty$ is said to be strongly singular if and only if its principal sections $A_n = [a_{ik}]_0^{n-1}$ are non-regular beginning with some $n = n_0$.

PROPOSITION 5.14. An infinite H-matrix is strongly singular if and only if it is a singular (∞,∞)-extension of a regular H-matrix. An infinite T-matrix is strongly singular if and only if it is a half-singular (∞,∞)-extension of a regular T-matrix.

6. T- AND H-MATRICES WITH NON-REGULAR PRINCIPAL SECTIONS

After the investigations of Section 5 we are able to extend the results of Section 3 to regular but not strongly regular T- and H-matrices. In particular, an inversion algorithm for arbitrary regular T- and H-matrices is constructed.

6.1. Inversion algorithm - Toeplitz case. Suppose $a = (a_i)_{-N}^{N-1} \in \mathbb{C}^{2N}$ and $T = T_N(a)$ is a regular T-matrix. We consider the principal sections $T_m(a) = [a_{i-k}]_0^{m-1}$ of T. Let $\{m_0,\dots,m_d\}$, $0 = m_0 < \dots < m_d = N$, denote the set of all indices m, for which $T_m(a)$ is regular. The idea of the algorithm is to jump from a regular section $T_{m_i}(a)$ to the next $T_{m_{i+1}}(a)$. More precisely speaking, we shall establish a rule how to compute the solutions of the fundamental equations (3.1) and (3.3) for $n = m_{i+1}$ with the help of the fundamental solutions for $n = m_i$.

Suppose the solutions of the fundamental equations (3.1), (3.3) are known for $n = m_i =: r$. The first question is how to decide whether $T_{r+1}(a)$ is regular or not. If $T_{r+1}(a)$ is regular then, according to Prop. 3.1,

$$\varepsilon_r := [a_r, a_{r-1}, \dots, a_0]\, \underline{z}^r \neq 0,$$

where $\underline{z}^r = [{z^r \atop -1}]$. The converse is also true, since in case $\varepsilon_r = 0$ the vector $\underline{z}^r$ is a non-trivial solution of the homogeneous equation $T_{r+1}(a)\, u = 0$.

If $T_{r+1}(a)$ is regular, then the recursion formulae (3.4), (3.5) can be applied. Otherwise, if $T_{r+1}(a)$ is not regular, the recursion algorithm consists of the following steps.

<u>1st step.</u> Determination of the smallest integer n, where $n > r$, such that $T_{n+1}(a)$ is regular $(n+1 = m_{i+1})$ and evaluation of the (ϱ, π_+, π_-)-characteristic of $T_n(a)$.

<u>2nd step.</u> Determination of a vector $z' \in \mathbb{C}^{n+1}$ such that $T_{n+1}(a)z'$ coincides with $g^{n+1} := (a_{i-n-1})_0^n$ except the first $\pi_- + 1$ and the last $\pi_+ + 1$ components.

<u>3rd step.</u> Evaluation of the solution x^{n+1} of (3.1) and correction of z' to yield the solution z^{n+1} of (3.3).

These three steps also must be applied if already $T_1(a) = a_0 = 0$.[1]

We explain now how to realize the steps of the recursion.

<u>1st step.</u> As remarked above in case of non-regular $T_{r+1}(a)$ there is

$$\ker T_{r+1}(a) = \text{lin}\{u\}, \quad u := [{z^r \atop -1}].$$

We introduce numbers

$$\sigma_i^+ := [a_{r+i} \dots a_i]u, \quad \sigma_i^- := [a_{-i} \dots a_{-r-i}]u$$

for $i = 0,1,\dots$ and

$$l = \min\{i : \sigma_i^+ \neq 0\}, \quad m := \min\{i : \sigma_i^- \neq 0\}.$$

<u>LEMMA 6.1.</u> The smallest number $n \geq r$, for which $T_{n+1}(a)$ is regular is determined by

$$n = r + l + m - 1. \tag{6.1}$$

---

[1] In this case we have to set $r := 0$, $u := -1$ and $z' := (0\dots0)$.

Furthermore,

$$\operatorname{char} T_n(a) = (r, m-1, l-1). \tag{6.2}$$

PROOF. Let n be defined by (6.1). Consider the vector $p_s = I^{m+s}_{n,r+1}u$ $(s = 0, \pm 1, \dots)$, which has vanishing first $m + s - 1$ and last $l - s - 1$ components. We have

$$T_n(a)p_s = (\sigma^-_{m+s-1}, \sigma^-_{m+s-2} \dots \sigma^+_{l-s-2}, \sigma^+_{l-s-1}).$$

Therefore, according to Theorem 5.16 and the definition of m, l, we obtain

$$\ker T_n(a) = \operatorname{lin}\{p_o\}.$$

In virtue of Prop. 5.13' this implies (6.2).
Since $T_n(a)$ is not half-regular and $\dim \ker T_n(a) = 1$, the matrix $T_{n+1}(a)$ must be regular. Furthermore, all sections $T_k(a)$ for $k = r+1, \dots, n$ are non-regular. Hence n is the smallest number, for which $T_{n+1}(a)$ is regular. ≡

2nd step. We introduce the vectors

$$g(k) := (a_{i-r-k})_{i=o}^{r-1} \in \mathbb{C}^r \qquad (k = 0, 1, \dots).$$

Then we have the recursion

$$g(0) = g^r \text{ (c.f. (3.3)) and } g(k+1) = U_r g(k) + a_{-r-k-1}e_o. \tag{6.3}$$

We consider the equations

$$T_r(a)z(k) = g(k) \qquad (k = 0, 1, \dots). \tag{6.4}$$

LEMMA 6.2. The solutions of (6.4) satisfy the following recursion formula

$$z(k+1) = U_r z(k) + \eta_k z(0) + \varepsilon_k x^r, \tag{6.4a}$$

where $\eta_k := e^T_{r-1}z(k)$, $\varepsilon_k := a_{-r-k-1} - \widehat{g(0)}^T z(k)$ and x^r is the solution of (3.1) for $n = r$.[1)]

PROOF. By formula (1.1) we have

$$T_r(a)U_r z(k) = T_r(a)z(k+1) - \varepsilon_k e_o - \eta_k g(0).$$

This implies (6.4a). ≡

Now we define

$$z' = I^l_{n+1,r} z(m).$$

Then $g' := T_{n+1}z'$ coincides with g^{n+1} with the exception of the first $l = \pi_- + 1$ and the last $m = \pi_+ + 1$ components.
Let us illustrate the idea of the second step in a picture.

---

1) As above $\hat{c}$ denotes the vector c with reverse order of components.

PICTURE 6.1.

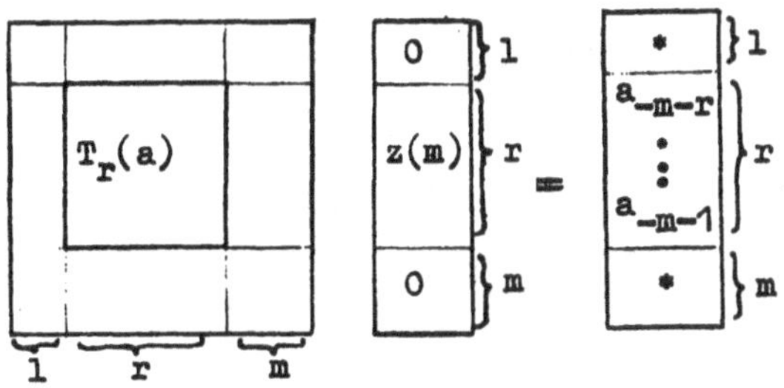

This picture illustrates the equation

$$T_{n+1}(a) \qquad z' \;=\; g' .$$

3rd step. Let P be the $(n+1)\times(n-r)$ matrix $[I^0u \ldots I^{m+l-1}u]$. Then we have by definition of the numbers $\sigma_i^{\pm}$

$$T_{n+1}(a)\,P = \left[\begin{array}{cccccc} 0 & \ldots & 0 & \sigma_m^- & \ldots & \sigma_{m+l-1}^- \\ \vdots & & \vdots & & \ddots & \vdots \\ 0 & \ldots & 0 & 0 & & \sigma_m^- \\ 0 & \ldots & & & \ldots & 0 \\ \vdots & & & & & \vdots \\ 0 & \ldots & & & \ldots & 0 \\ \sigma_l^+ & 0 & & 0 & \ldots & 0 \\ \vdots & \ddots & & \vdots & & \vdots \\ \sigma_{l+m-1}^+ & \ldots & \sigma_l^+ & 0 & \ldots & 0 \end{array}\right] \begin{array}{l} \left.\vphantom{\begin{array}{c}0\\0\\0\end{array}}\right\} l \\ \left.\vphantom{\begin{array}{c}0\\0\\0\end{array}}\right\} r \\ \left.\vphantom{\begin{array}{c}0\\0\\0\end{array}}\right\} m \end{array} . \tag{6.5}$$

From this follows particularly for the solution x^{n+1} of the fundamental equation $T_{n+1}x^{n+1} = e_0$

$$x^{n+1} = \frac{1}{\sigma_m^-} I^m u . \tag{6.6}$$

Furthermore, the equations $T_{n+1}(a)x_k = e_k$ for $k = 0,\ldots,l-1$ and $k = n-m+1,\ldots,n$ easily can be solved and their solutions x_k can be used to correct the vector z' to yield the solution z^{n+1} of the second fundamental equation

$$T_{n+1}(a)z^{n+1} = g^{n+1} . \tag{6.7}$$

This correction can be done successively in the following way. We put

$$z'(0) := z', \quad z'(k+1) := z'(k) + \gamma_k^- I^{m+l-1-k}u, \tag{6.8}$$

where

$$\gamma_k^- := \frac{1}{\sigma_m^-}(a_{l-k-n-2} - [a_{l-1-k} \dots a_{l-1-k-n}]z'(k)) \ (k=0,1,\dots,l-1).$$

The vector $T_{n+1}(a)\ z'(k+1)$ coincides with g^{n+1} in the same components as $T_{n+1}(a)\ z'(k)$ and, additionally, in the $(l+1-k)$th component. In particular, for $z'' := z'(l)$ the vector $T_{n+1}z''$ coincides with g^{n+1} with the exception of the last m components. These components we correct successively by means of the recursion

$$z''(0) := z'', \quad z''(k+1) := z''(k) + \gamma_k^+ I^k u, \tag{6.9}$$

where

$$\gamma_k^+ := \frac{1}{\sigma_l^+}(a_{k-m} - [a_{n+k-m+1} \dots a_{k-m+1}]z''(k)) \ (k=0,1,\dots,m-1).$$

Finally we obtain that $z''(m)$ is just the fundamental solution z^{n+1} of (6.7).
The steps of our algorithm are completely realized.

<u>REMARK 6.1.</u> It is easily seen that the algorithm requires only $O(N^2)$ operations, as in the strongly regular case.

<u>6.2. Inversion algorithm - Hankel case.</u> The algorithm for H-matrix inversion follows the same idea as that one for T-matrix inversion, namely, the jumping from a regular section to the next larger regular section. Let $H_r(b)$ be a regular section of a regular H-matrix $H_N(b)$, $b = (b_i)_0^{2N-1}$, and y^r, w^r the solutions of the fundamental equations (3.15), (3.16) for $n = r$. The evaluation of the corresponding fundamental solutions of the next regular section consists of the same steps as in the Toeplitz case.

<u>1st step.</u> Determination of the smallest $n \geq r$ for which $H_{n+1}(b)$ is regular. We have $n = r$ if and only if $\varepsilon := [b_{r+1} \dots b_{2r+1}]\ u \neq 0$, where $u = \begin{bmatrix} w^r \\ -1 \end{bmatrix}$. In case $\varepsilon = 0$ there is $\ker H_{r+1}(b) = \mathrm{lin}\{u\}$.
We define

$$\sigma_i := [b_{r+i} \dots b_{2r+i}]u, \quad m := \min\{i : \sigma_i \neq 0\}.$$

<u>LEMMA 6.1'.</u> The smallest number $n \geq r$ for which $H_{n+1}(b)$ is regular is determined by

$$n = r + m. \tag{6.10}$$

Furthermore,

$$\mathrm{char}\ H_n(a) = (r, m-1). \tag{6.11}$$

<u>PROOF.</u> If n is defined by (6.10), then
$$\ker H_n(b) = \mathrm{lin}\{I^0 u\},$$

and the assertion follows using the same arguments as in Lemma 6.1. $\equiv$

2nd step. We introduce the vectors

$$h(k) := (b_{r+k+i})_{i=0}^{r-1} \quad (k = 0,1,\dots).$$

Then we have

$$h(0)=h^r \quad \text{and} \quad h(k+1) = U_r^T h(k) + b_{2r+k}.$$

To establish a recursion formula for the solutions of $H_r(b)w(k) = h(k)$ we observe (see(1.19))

$$H_r(b)U_r w(k) = U_r^T h(k) + (h(0)^T w(k))e_{r-1} - (e_{r-1}^T w(k))h(0).$$

This implies

$$w(k+1) = U_r w(k) + \eta_k w(0) + \varepsilon_k x^k, \tag{6.12}$$

where $\eta_k := e_{r-1}^T w(k)$, $\varepsilon_k := b_{2r+k} - h(0)^T w(k)$ $(k = 0,\dots,m-1)$. Define $w' := I_{n+1,r}^0 w(m)$, then $H_{n+1}w'$ coincides with h^{n+1} with the exception of the last $m + 1$ components.

3rd step. By definition of the numbers σ_i we have

$$H_{n+1}\ I_{n+1,r}^k u = (0\dots0,\ \sigma_m\dots\sigma_{m+k}) \quad (k = 0,\dots,m). \tag{6.13}$$

We define vectors $w'(k)$

$$w'(0) := w',\quad w'(k+1) := w'(k) + \gamma_k\ I^{m-1-k}u,$$

where

$$\gamma_k := \frac{1}{\sigma_m}(b_{r+k+n+2} - [b_{r+k+1}\dots b_{r+k+n+1}]w'(k)).$$

Then the recursion from k to $k + 1$ corrects the $(n-m+k)$th component of the right side. Therefore $w^{n+1} := w'(m+1)$ is just the fundamental solution of (3.16) for $H_{n+1}(b)$. The other fundamental solution y^{n+1} of (3.15) is obviously given by

$$y^{n+1} := \frac{1}{\sigma_m} I^0 u.$$

6.3. Generalized LU-factorization. We recall that a regular matrix admits an LU-factorization if and only if it is strongly regular. We shall show in this subsection that arbitrary H- and T-matrices admit an LU-factorization in a generalized sense. We make use of the designations introduced in Subsection 3.6.

DEFINITION 6.1. Two $n\times n$ matrices A and B will be called LU-equivalent if there are matrices $P_\pm \in \mathcal{L}_\pm^n$ such that $B = P_- A P_+$. Obviously, the LU-equivalence is an equivalence relation.

THEOREM 6.1. Any H-matrix H is LU-equivalent to a block-diagonal matrix

$$D = \mathrm{diag}(\gamma_1 J_{\delta_1} \dots \gamma_{d-1} J_{\delta_{d-1}}, \underbrace{0 \dots 0}_{\alpha}, \gamma_d J_{\delta_d}),^{1)} \tag{6.14}$$

where the numbers γ_i, δ_i $(i = 1,\dots,d)$ are uniquely determined by H. More in detail, suppose $H = H_n(b)$, $b = (b_i)_0^{2n-2}$, $m_0 := 0$, $m_d := n$ and let $m_i (i=1,\dots,d-1)$, $0 < m_1 < \dots < m_{d-1} < n$, denote the indices $k < n$, for which $H_k(b)$ is regular. Then $\alpha = \dim \ker H$,

$$\delta_i = m_i - m_{i-1} \quad (i = 1,\dots,d-1), \; \delta_d = n - m_{d-1} - \alpha \tag{6.14a}$$

$$\gamma_i = [b_{m_i-1} \dots b_{m_i+m_{i-1}-1}] \begin{bmatrix} -w^{m_{i-1}} \\ 1 \end{bmatrix} \neq 0, \tag{6.14b}$$

where w^k $(k = m_i)$ denotes the solution of the fundamental equation (3.16). Furthermore, suppose $H = H_- D H_+$, where $H_\pm \in \mathcal{L}_\pm^n$ and D is given by (6.14). Then $H_\pm$ can be chosen such that $H_- = H_+^T$.

PROOF. For short we denote $m := m_{d-1}$. Furthermore, we put

$$p := \begin{bmatrix} -w^m \\ 1 \end{bmatrix}$$

and introduce the nxn matrix

$$P = [e_0 \dots e_{m-1} \; I^0 p \dots I^{n-m-1} p] \in \mathcal{L}_+^n.$$

Then it is easily verified that

$$P^T H_n(b) P = \begin{bmatrix} H_m(b) & 0 \\ 0 & \Gamma \end{bmatrix},$$

where Γ has the form

$$\Gamma = \begin{bmatrix} 0 & 0 \\ 0 & \Gamma_0 \end{bmatrix} \begin{matrix} \}\alpha \\ \}\pi \end{matrix}, \qquad \Gamma_0 = \begin{bmatrix} & & g_0 \\ & \cdot^{\cdot^{\cdot}} & \vdots \\ g_0 & \cdots & g_{\pi-1} \end{bmatrix}$$

with $g_0 = \gamma_d \neq 0$ and $\pi = n - \alpha - m = \delta_d$. Let σ_i be the numbers defined by

$$\sigma_0 + \sigma_1 \lambda + \dots = \sqrt{1 + \frac{g_1}{g_0} \lambda + \dots + \frac{g_{\pi-1}}{g_0} \lambda^{\pi-1}} \; .$$

Then we have

$$\Gamma_0 = g_0 \Sigma^T J_\pi \Sigma,$$

where

$$\Sigma = \begin{bmatrix} \sigma_0 & \cdots & \sigma_{\pi-1} \\ & \ddots & \vdots \\ & & \sigma_0 \end{bmatrix} \in \mathcal{L}_+^\pi \; .$$

This implies

$$P^T \begin{bmatrix} I_{n-\pi} & 0 \\ 0 & \Sigma^T \end{bmatrix} H_n(b) \begin{bmatrix} I_{n-\pi} & 0 \\ 0 & \Sigma \end{bmatrix} P = \begin{bmatrix} H_m(b) & 0 \\ 0 & 0 \\ 0 & \gamma J_\pi \end{bmatrix} \} \alpha \; .$$

---

1) The matrices J_k are introduced in Subsection 0.2.

We repeat now this construction replacing $H_n(b)$ by $H_m(b)$ and $H_m(b)$ by $H_{m_{d-2}}(b)$ and so on. After d steps we obtain a representation

$$\tilde{P}^T H_n(b) \tilde{P} = D,$$

where D is of the form (6.14) and $\tilde{P} \in \mathcal{L}_+^n$. Thus the LU-equivalence of H and D is proved. The uniqueness of D is easily verified. By our construction we also proved the last assertion of the theorem. ≡

<u>THEOREM 6.2.</u> Any T-matrix T is LU-equivalent to a block-diagonal matrix

$$D = \mathrm{diag}(\Delta_1 \ldots \Delta_{d-1}, \Delta_d), \tag{6.15}$$

where

$$\Delta_i = \begin{bmatrix} 0 & \beta_i I_{\delta_i} \\ \gamma_i I_{\varepsilon_i} & 0 \end{bmatrix} \quad (i = 1,\ldots,d-1), \tag{6.15a}$$

$$\Delta_d = \begin{bmatrix} 0 & 0 & \beta_d I_{\delta_d} \\ 0 & 0 & 0 \\ \gamma_d I_{\varepsilon_d} & 0 & 0 \end{bmatrix} \}\alpha \quad , \tag{6.15b}$$

$\delta_i, \varepsilon_i > 0$ or $\delta_i = 0$, $\varepsilon_i = 1$, and all occurring numbers are uniquely determined by T. More in detail, suppose $T = T_n(a)$, $a = (a_i)_{1-n}^{n-1}$, let m_i $(i = 1,\ldots,d-1)$ $(m_o := 0 < m_1 < \ldots < m_{d-1} < m_d := n)$ denote the indices, for which $T_k(a)$ is regular and

$$\mathrm{char}\ T_{m_i-1}(a) = (m_{i-1}, \pi_i^+, \pi_i^-).$$

Then

$$\varepsilon_i = \pi_i^+ + 1,\ \alpha = n-\pi_d^+-\pi_d^--m_{d-1},\ \delta_i = \begin{cases} \pi_i^-+1 & : \pi_i^- > 0 \\ 0 & : \pi_i^- = 0 \end{cases},$$

$$\gamma_i = [a_{n-\varepsilon_i-1} \ldots a_{n-\varepsilon_i-1-m_{i-1}}]\begin{bmatrix} -z^{m_{i-1}} \\ 1 \end{bmatrix}, \tag{6.16a}$$

$$\beta_i = -\frac{1}{\zeta_{i-1}}[a_{m_{i-1}-n+\delta_i-1} \ldots a_{-n+\delta_i-1}]\begin{bmatrix} -z^{m_{i-1}} \\ 1 \end{bmatrix}, \tag{6.16b}$$

where z^k denotes the solution of the fundamental equation (3.3) for $T_k(a)$ and ζ_i is the first component of z^{m_i}.

<u>PROOF.</u> For short we denote $m := m_{d-1}$. Furthermore, we put

$$p := \begin{bmatrix} -z^m \\ 1 \end{bmatrix},\ \zeta := \zeta_{d-1}\ (\neq 0),\ \tilde{p} := -\frac{1}{\zeta} J_{m+1}p \tag{6.17}$$

and introduce the matrices

$$P = [e_0 \dots e_{m-1}\ I^0 p \dots I^{n-m-1} p] \in \mathcal{L}_+^n ,$$

$$Q = [e_0 \dots e_{m-1}\ I^0 \tilde{p} \dots I^{n-m-1} \tilde{p}] \in \mathcal{L}_+^n .$$

We obtain after elementary computation

$$Q^T\, T_n(a)\, P = \begin{bmatrix} T_m(a) & 0 \\ 0 & \Delta \end{bmatrix} ,$$

where

$$\Delta = \begin{bmatrix} 0 & 0 & \beta_d \Delta_- \\ 0 & 0 & 0 \\ \gamma_d \Delta_+ & 0 & 0 \end{bmatrix} \begin{matrix} \} \pi_- \\ \} \alpha \\ \} \pi_+ \end{matrix}$$

for some $\Delta_+ \in \mathcal{L}_-^{\pi_+}$ and $\Delta_- \in \mathcal{L}_-^{\pi_-}$. Then

$$\Delta = \Sigma_-\, \Delta_d\, \Sigma_+ ,$$

where Δ_d is defined by (6.15b) and

$$\Sigma_- = \begin{bmatrix} I_{n-1} & 0 \\ 0 & \Delta_+ \end{bmatrix}, \quad \Sigma_+ = \begin{bmatrix} I_{n-m} & 0 \\ 0 & \Delta_- \end{bmatrix}.$$

This implies

$$Q^T \begin{bmatrix} I_{n-\pi_+} & 0 \\ 0 & \Delta_+ \end{bmatrix} T_n(b) \begin{bmatrix} I_{n-\pi_-} & 0 \\ 0 & \Delta_- \end{bmatrix} P = \begin{bmatrix} T_m(a) & 0 \\ 0 & \Delta_d \end{bmatrix}.$$

Replacing now $T_n(a)$ by $T_m(a)$ and $T_m(a)$ by $T_{m_{d-2}}(a)$ and so on and repeating the construction we obtain after d steps a representation

$$\tilde{Q}^T\, T_n(a)\, \tilde{P} = D,$$

where D has the form (6.15) and $\tilde{Q}, \tilde{P} \in \mathcal{L}_+^n$. The LU-equivalence of $T_n(a)$ and D is proved. The uniqueness of D is easily verified. ≡

Let us investigate now the specific of Hermitian T-matrices. In this case the vector p defined by (6.17) is symmetric, i.e. $J_{m+1}p = \overline{p}$. Therefore, we have $\zeta_i = 1$ $(i = 1,\dots,d)$ and $Q^T = P^*$, $\tilde{Q}^T = \tilde{P}^*$ in the proof of Theorem 6.2. Moreover, by construction described in this proof, it is obvious that $\delta_i = \varepsilon_i$ and $\beta_i = \overline{\gamma}_i$. These facts lead to the following assertion.

COROLLARY 6.1. Any Hermitian T-matrix $T \in \mathcal{T}(n)$ can be represented in the form

$$T = T_+^*\, D\, T_+ ,$$

where $T_+ \in \mathcal{L}_+^n$ and D is of the form (6.15) with $\varepsilon_i = \delta_i$, $\beta_i = \overline{\gamma}_i$ if $\delta_i \neq 0$ and $\gamma_i \in \mathbb{R}$ if $\delta_i = 0$.

REMARK 6.2. Replacing the LU-equivalence by the UL-equivalence theorems analogously to Theorem 6.1 and 6.2 can be formulated. In the Toeplitz case (not in the Hankel case) the block-diagonal matrix of UL-equivalence is the transposed of the matrix D defined by (6.15).

THEOREM 6.3. Any $m\times n$ H-Bezoutian B is UL-equivalent to a block-diagonal matrix of the form

$$D = \operatorname{diag}(O_{\alpha\beta}, \gamma_1 J_{\delta_1} \dots \gamma_d J_{\delta_d}), \tag{6.18}$$

where $O_{\alpha\beta}$ is the $\alpha\times\beta$ null matrix ($\alpha = \dim\ker B$, $\beta = \dim\ker B^T$) and the numbers δ_i and $\gamma_i \neq 0$ are uniquely determined by B. In the particular case that B is a classical $n\times n$ Bezoutian there is a representation of the form $B = B_-^T D B_-$, where $B_- \in \mathcal{L}_-^n$.

PROOF. By Prop. 2.6 B can be represented in the form

$$B = [I^o_{m,m-r} \quad D_r(p)^T]\begin{bmatrix} O & O \\ O & B' \end{bmatrix}\begin{bmatrix} I^o_{n-r,n} \\ D_r(q) \end{bmatrix}, \tag{6.19}$$

where B' is a regular $r\times r$ Bezoutian and $p(\lambda)$, $q(\lambda)$ are polynomials with non-zero highest order coefficients. Since the polynomials $p(\lambda)$ and $q(\lambda)$ may assumed to be monic the right and left factor in (6.19) belongs to $\mathcal{L}_-^n$, $\mathcal{L}_+^m$, respectively. Hence B is UL-equivalent to $\begin{bmatrix} O & O \\ O & B' \end{bmatrix}$. By Cor. 2.1 $(B')^{-1}$ is an H-matrix. Applying Theorem 6.1 we obtain the UL-equivalence of B and (6.18). To prove the last assertion it is sufficient to note that in the case of a classical Bezoutian B we have $p = q$. ≡

COROLLARY 6.2. Suppose $B = [b_{ik}]_0^{m-1}\,{}_0^{n-1}$ is an H-Bezoutian and rank $B = r$. Then the matrix

$$[b_{ik}]_{m-r}^{m-1}\,{}_{n-r}^{n-1}$$

is regular.

Finally, let us remark that the block diagonal matrix, which is UL-equivalent to an H-Bezoutian has the form

$$\operatorname{diag}(O_{\alpha_1\beta_1}, \gamma_1 J_{\delta_1} \dots \gamma_d J_{\delta_d}\; O_{\alpha_2\beta_2}).$$

6.4. Evaluation of the signature. The algorithms described in Subsections 6.1, 6.2 can be used for the evaluation of the signature of arbitrary H- and T-matrices. Indeed, applying Theorem 6.1, Cor. 6.1 and Sylvester's inertia law we obtain the following rules.

PROPOSITION 6.1. For the signature of a real H-matrix $H_n(b)$ the following relation holds true

$$\operatorname{sgn} H_n(b) = \sum_{\delta_i \text{ odd}} \operatorname{sgn} \gamma_i,$$

where δ_i, γ_i are defined by (6.14a), (6.14b).

For the proof it is sufficient to note that

$$\operatorname{sgn} J_k = \begin{cases} 0 & \text{if } k \text{ is even} \\ 1 & \text{if } k \text{ is odd.} \end{cases}$$ [1] $\equiv$

PROPOSITION 6.2. For the signature of an Hermitian T-matrix $T_n(a)$ the following relation holds true

$$\operatorname{sgn} T_n(a) = \sum_{\delta_i=0} \operatorname{sgn} \gamma_i .$$

The proof follows from Cor. 6.1, Sylvester's law and the fact that

$$\operatorname{sgn} \begin{bmatrix} 0 & \bar{\gamma} I_\delta \\ \gamma I_\delta & 0 \end{bmatrix} = 0. \quad \equiv$$

Finally, let us note that the algorithms described in 6.1 and 6.2 clearly can be used for the rank determination of H- and T-matrices, since (c.f. Table 5.1)

$$\operatorname{rank} H_n(b) = \rho + \pi, \quad \operatorname{rank} T_n(a) = \rho + \pi_+ + \pi_- .$$

Using this and Prop. 6.1, 6.2 the inertia of H- and T-matrices is easily computed.

7. GENERALIZED INVERSES OF H-MATRICES

7.1. Two approaches for generalized inversion.

DEFINITION 7.1. An $n \times m$ matrix B is called generalized inverse to the $m \times n$ matrix A if and only if

$$ABA = A \quad \text{and} \quad BAB = B.$$

If, additionally, BA and AB are Hermitian matrices B is called the Moore-Penrose inverse or pseudoinverse of A.

It is well-known that any matrix has at least one generalized inverse, which is uniquely determined if and only if A is regular. More in detail, to any direct complement C_1 of $\ker A$ and C_2 of $\operatorname{im} A$ there exists exactly one generalized inverse B of A such that $\ker B = C_2$ and $\operatorname{im} B = C_1$. If one takes for C_1 the orthogonal complement of $\ker A$ and for C_2 the orthogonal complement of $\operatorname{im} A$ one obtains just the Moore-Penrose inverse and vice versa. Therefore, any matrix has exactly one Moore-Penrose inverse. Let us still note, that, obviously, for every generalized inverse B of A the operator AB is a projection onto $\operatorname{im} A$ and $I - BA$ is a projection onto $\ker A$.

In this section we deal with the problem of generalized inversion of H-matrices. We are interested in generalized inverses with special

---

[1] See also Cor. 8.4.

structure. We shall offer two variants for solving this problem. First we give a complete description of generalized inverses of H-matrices, which are Bezoutians. In our second approach we look for generalized inverses, which can be expressed by means of characteristic polynomials.

Both approaches base upon two principles of generalized inversion, which we shall present in this subsection

1. generalized inversion by restriction
2. generalized inversion by extension.

Let us begin with explaining the first principle. Suppose A is an $m\times n$ matrix with rank r, P is an $n\times r$ matrix and Q an $r\times m$ matrix. Further, we assume that P and Q satisfy the following conditions:

$$\operatorname{im} P \cap \ker A = \{0\}, \quad \ker P = \{0\}, \tag{7.1a}$$

$$\operatorname{im} A \cap \ker Q = \{0\}, \quad \operatorname{im} Q = \mathbb{C}^r. \tag{7.1b}$$

<u>LEMMA 7.1.</u> Let A, P, Q be matrices satisfying (7.1), then the matrix

$$A_0 := QAP$$

is regular and $B := PA_0^{-1}Q$ is generalized inverse to A. Moreover, any generalized inverse can be represented in this form.

<u>PROOF.</u> Obviously, $\operatorname{im} P$ is a direct complement of $\ker A$. Hence $\ker AP = \{0\}$ and $\operatorname{im} AP = \operatorname{im} A$. Analogously, $\operatorname{im} A$ is a direct complement of $\ker Q$. Therefore, $\ker A_0$ is trivial and $\operatorname{im} A_0 = \operatorname{im} Q = \mathbb{C}^r$, which means that A_0 is regular. Furthermore, we have

$$BAB = PA_0^{-1}QAPA_0^{-1}Q = PA_0^{-1}Q = B.$$

On the other side, there is

$$QABAP = QAP,$$

which means that QAB and Q coincide on $\operatorname{im} AP$. Since $\operatorname{im} AP = \operatorname{im} A$ this is the same as $QABA = QA$. From the latter equality we conclude that $\operatorname{im}(BA-I)$ has only the zero intersection with $\ker QA$.
Since $\ker QA = \ker A$, according to (7.1), this yields $ABA - A = 0$. Hence B is generalized inverse to A.
Now let B be an arbitrary generalized inverse of A and $\operatorname{rank} A = r$. Further, let P be an $n\times r$ matrix such that $\operatorname{im} P = \operatorname{im} B$ and Q be an $r\times m$ matrix such that $\ker Q = \ker B$. Then, according to the first part of our proof

$$\tilde{B} := Q(QAP)^{-1}P$$

is generalized inverse to A, too.
Since, moreover, $\ker \tilde{B} = \ker B$ and $\operatorname{im} \tilde{B} = \operatorname{im} B$ we conclude $B = \tilde{B}$, and the lemma is completely proved. ≡

Our second approach for generalized inversion bases upon the idea of extension of a singular matrix to a regular one. Let A be an $m\times n$ matrix with rank r and A_1, A_{-1}, A_0 matrices of suitable size such that the matrix of order $m + n - r$

$$\mathfrak{X} = \begin{bmatrix} A & A_{-1} \\ A_1 & A_0 \end{bmatrix}$$

is regular. Moreover, we assume that the following conditions are satisfied

$$\ker[A\ A_{-1}] = \{(x,0) : x \in \ker A\}, \tag{7.2a}$$

$$\ker[A^T A_{-1}^T] = \{(x,0) : x \in \ker A^T\}. \tag{7.2b}$$

It is easily seen that such matrices A_i $(i = 0, \pm 1)$ exist.

LEMMA 7.2. Assume that A_i $(i = 0, \pm 1)$ satisfy the conditions (7.2) and

$$\mathfrak{X}^{-1} = \begin{bmatrix} B & B_{-1} \\ B_1 & B_0 \end{bmatrix},$$

where B is an $n\times m$ matrix. Then B is generalized inverse to A.

PROOF. We have, by definition,

$$[A\ A_{-1}]\begin{bmatrix} B \\ B_1 \end{bmatrix} = I_m,$$

hence

$$[A\ A_{-1}]\begin{bmatrix} B\ Ax - x \\ B_1\ Ax \end{bmatrix} = 0$$

for all $x \in \mathbb{C}^n$. According to (7.2a) we conclude $B_1\ Ax = 0$. This yields $ABAx = Ax$. The equality $BAB = B$ we obtain by changing the role of A_i and B_i, but for this we have to show that (7.2b) implies

$$\ker[B\ B_{-1}] = \{(x,0) : x \in \ker B\}. \tag{7.3}$$

Indeed, (7.2b) is equivalent to

$$\operatorname{im}\begin{bmatrix} A \\ A_1 \end{bmatrix} = \{\begin{bmatrix} Ax \\ y \end{bmatrix} : x \in \mathbb{C}^n,\ y \in \mathbb{C}^{m-r}\}.$$

Furthermore, we have

$$[B\ B_{-1}]\begin{bmatrix} A \\ A_1 \end{bmatrix} = I_n,$$

which means that $\operatorname{im}\begin{bmatrix} A \\ A_1 \end{bmatrix}$ is a direct complement of $\ker[B\ B_{-1}]$. This yields (7.3), and so we proved that B is generalized inverse to A. ≡

REMARK 7.1. Any generalized inverse can be obtained by the inversion of an extended matrix.

7.2. Bezoutians as generalized inverses. In this subsection we give a complete description of all generalized inverses of H-matrices, which are Bezoutians or classical Bezoutians. For this we use the restriction variant of generalized inversion and the following observation: If H is an $m\times n$ H-matrix, $D_{r_1}(a)$ and $D_{r_2}(b)$ are difference operators (see 0.3) of suitable size, then the matrix $D_{r_1}(b) H D_{r_2}(a)^T$ is an H-matrix again.

Before we shall formulate the basic theorem of this subsection let us introduce some notations. Let H be a non-regular $n\times n$ H-matrix with rank r. We denote by $p(\lambda) \in \mathbb{C}^{r+1}(\lambda)$ a characteristic polynomial of H corresponding to the lowest characteristic degree (or, with other words, $\ker H$ is the linear hull of the U-chain generated by p).

THEOREM 7.1. Let H be a non-regular $n\times n$ H-matrix with rank r and $d(\lambda) \in \mathbb{C}^{n-r+1}(\lambda)$ a polynomial, which has no common zero (inclusively ∞) with $p(\lambda)$. Then the H-matrix

$$H_0 := D_r(d)\; H\; D_r(d)^T$$

is regular, and

$$B := \mathrm{Bez}(a(\lambda)\; d(\lambda),\; b(\lambda)\; d(\lambda)) \tag{7.4}$$

is a generalized inverse of H, where

$$H_0^{-1} = \mathrm{Bez}(a,b).$$

On the other hand, any generalized inverse of H, which is a classical Bezoutian, can be constructed in this way.

PROOF. The image of $D_r(d)^T$ is just the linear hull of the U-chain generated by d. Therefore, the condition

$$\ker H \cap \mathrm{im}\; D_r(d)^T = \{0\}$$

is fulfilled if and only if the system of the two U-chains generated by d and p is linearly independent. This is equivalent to the regularity of the matrix $\mathrm{Res}(d,p)$ and, by Cor. 2.5, equivalent to $\sigma(d) \cap \sigma(p) = \emptyset$. Analogously we obtain

$$\mathrm{im}\; H \cap \ker D_r(d) = \{0\}.$$

Therefore, the assumptions of Lemma 7.1 are satisfied. We obtain that H_0 is regular and $B := D_r(d)^T H_0^{-1} D_r(d)$ is a generalized inverse of H. According to Prop. 2.6 there is $B = \mathrm{Bez}(ad,bd)$, where $H_0^{-1} = \mathrm{Bez}(a,b)$. The first part of the theorem is proved.

Now let $B = \mathrm{Bez}(\tilde{a},\tilde{b})$ be a generalized inverse of H, $d(\lambda)$ be the g.c.d. of $\tilde{a}(\lambda)$ and $\tilde{b}(\lambda)$, $\tilde{a}(\lambda) = a(\lambda)d(\lambda)$, $\tilde{b}(\lambda) = b(\lambda)d(\lambda)$. Then, by Prop. 2.6,

$$B = D_r(d)^T\; \mathrm{Bez}(a,b)\; D_r(d),$$

where $B_0 := \mathrm{Bez}(a,b)$ is a regular $r\times r$ matrix. By assumption we have $BHB = B$, which is

$$D_r^T\; B_0\; H_0\; B_0\; D_r = D_r^T\; B_0\; D_r,$$

where $H_0 := D_r H D_r^T$ and $D_r := D_r(d)$. This implies $B_0 H_0 B_0 = B_0$. Since B_0 is regular we conclude $B_0 = H_0^{-1}$. In that manner we proved that any generalized inverse of H, which is a classical Bezoutian has the form $B = \mathrm{Bez}(ad,bd)$, where $\mathrm{Bez}(a,b) = (D_p(d)\ H\ D_p(d)^T)^{-1}$. It remains to show that $d(\lambda)$ and $p(\lambda)$ have no common root. Suppose they have a common root. Then $\mathrm{im}\ D_r^T$ and $\ker H$ have a non-trivial intersection. Obviously, it follows that $\ker H_0$ is non-trivial, which is a contradiction. The theorem is proved. ≡

REMARK 7.2. Suppose $H = H_n(b)$, then $D_r(d)\ H\ D_r(d)^T = H_r(\tilde{b})$, where $\tilde{b} = D_{2r-1}(d)b$.

From Remark 7.2 one may conclude, in particular, that the computation of H_0 and, therefore, the generalized inversion by Theorem 7.1 requires only $O(n^2)$ operations.

REMARK 7.3. For the matrix B defined by (7.4) the following relation holds true

$$\ker B = \ker B^T = \mathrm{lin}\{1^k(\lambda_i) : k=0,\ldots,r_i-1;\ \lambda_i \in \sigma(d)\};^{1)}$$

where r_i is the multiplicity of the root λ_i.

From this fact we conclude that to any system of n-r numbers ($\in \mathbb{C}$), which are no roots of $p(\lambda)$, corresponds exactly one generalized inverse of H, which is a classical Bezoutian and vice versa.

Now let us consider rectangular H-matrices H and look for general Bezoutians as generalized inverses of H.

THEOREM 7.2. Let H be an m×n H-matrix with rank $r < \min\{m,n\}$ and $p(\lambda)$ a characteristic polynomial of H with lowest characteristic degree. Further, let $d_1(\lambda) \in \mathbb{C}^{m-r+1}(\lambda)$, $d_2(\lambda) \in \mathbb{C}^{n-r+1}(\lambda)$ be polynomials, which have no common roots with $p(\lambda)$. Then the H-matrix

$$H_0 := D_r(d_1)\ H\ D_r(d_2)^T$$

is regular and

$$B := \mathrm{Bez}(ad_2, ad_1, bd_2, bd_1)$$

is generalized inverse to H, where

$$H_0^{-1} = \mathrm{Bez}(a,b).$$

On the other side, any generalized inverse of H, which is a Bezoutian can be constructed in this way.

We omit the proof of this theorem, because it is similar to that one of Theorem 7.1.

---

[1] For the definition of $1^k(\lambda)$ see 0.3.

It remains to consider rectangular H-matrices of maximal rank or, which is the same, H-matrices having a one side inverse. Without loss of generality we restrict ourselves to the case $m < n$.

<u>THEOREM 7.3.</u> Let H be an $m\times n$ H-matrix with $\operatorname{rank} H = m$ and let $d(\lambda) \in \mathbb{C}^{n-m+1}(\lambda)$ be a polynomial such that

$$\ker H \cap \operatorname{im} D_m(d) = \{0\}. \tag{7.5}$$

Then $H_o := H\, D_m(d)$ is regular and

$$B := \operatorname{Bez}(ad, a, bd, b)$$

is right inverse to H. On the other side, any right inverse of H, which is a Bezoutian, can be constructed in this way.

The proof is analogous to that one of Theorem 7.1.
Let us note that in the maximal rank case we have no "spectral" characterization of the condition (7.5), but we can formulate this condition in the following form.

<u>REMARK 7.4.</u> Let $p(\lambda)$ and $q(\lambda)$ be characteristic polynomials of H, then $\ker H$ is the linear hull of two U-chains

$$I^o p, \dots, I^{k-1} p \quad \text{and} \quad I^o q, \dots, I^{l-1} q.$$

The condition (7.5) is satisfied if and only if the matrix

$$[D_k(p)^T \quad D_l(q)^T \quad D_m(d)^T]$$

is regular.
Another approach of one-side inversion will be given in Subsection 7.5.

<u>7.3. Construction of generalized inverses with the help of characteristic polynomials.</u>

As shown above the inverse of a regular H-matrix H is the Bezoutian of two characteristic polynomials of H. In this subsection we shall see that some generalized inverses of non-regular H-matrices can be expressed by means of the characteristic polynomials. Before formulating the corresponding theorem let us note one circumstance concerning the characteristic polynomials. Let $p(\lambda)$ and $q(\lambda)$ be characteristic polynomials of H with characteristic degrees n_1 and n_2, where $n_1 \le n_2$. By Prop. 5.5 $p(\lambda)$ and $q(\lambda)$ have no common root, in particular ∞ is no common root. That means one of the polynomials has a non-zero highest order coefficient. Since $q(\lambda)$ is only unique up to a linear combination of $p(\lambda), \dots, \lambda^{n_2-n_1} p(\lambda)$ we may assume that $q(\lambda)$ has a non-zero highest order coefficient and, moreover, that $q(\lambda)$ is monic.

<u>THEOREM 7.4.</u> Let $H = H_{mn}(b)$, $b = (b_i)_o^{m+n-2}$, be an H-matrix with $r = \operatorname{rank} H < \min\{m,n\}$ and let $p(\lambda)$, $q(\lambda)$ be characteristic poly-

nomials of H with characteristic degrees n_1, n_2, where $q(\lambda)$ is monic. Then

$$B = \frac{1}{\varepsilon} I^o_{nl} \operatorname{Bez}(q,p) I^o_{lm} \text{ [1)]} \quad (l = m+n-r) \tag{7.6}$$

is generalized inverse to H, where $\varepsilon := [b_{l-1} \dots b_{l+r-1}]\, p \neq 0$.

PROOF. According to Table 5.1 it is obvious that $n_1 = r$, $n_2 = m+n-r = l$.
We define numbers $b_{m+n-1}, \dots, b_{2l-1}$ recursively by

$$b_k := [b_{k-l} \dots b_{k-1}] q^o \quad (k = m+n-1, \dots, 2l-1),$$

where $q^o(\lambda) := \lambda^n - q(\lambda)$, and consider the matrix $H_l(\tilde{b})$ with $\tilde{b} = (b_i)_o^{2l-2}$. Then we have

$$H_l(\tilde{b}) q^o = (b_{l+i})_{i=o}^{l-1}, \tag{7.7}$$

$$H_l(\tilde{b}) [\begin{smallmatrix} p \\ 0 \end{smallmatrix}] = \varepsilon\, e_{l-1}. \tag{7.8}$$

Let us show that $\varepsilon \neq 0$. For this we suppose for a moment $\varepsilon = 0$ and consider the family of H-matrices $H^k := H_{m+n+1-k,k}(\bar{b})$, where $\bar{b} := (b, b_{m+n-1})$. Then the matrix H^{r+1} has the partial indices $(1, -l+r)$ and, therefore, H^{l+1} has the partial indices $(l-r+1, 0)$. The latter means that the kernel of H^{l-1} is spanned by a U-chain generated by p. On the other hand we have, by definition, $q \in \ker H^{l-1}$, hence q must be a linear combination of this U-chain, which contradicts the definition of q. This contradiction yields $\varepsilon \neq 0$.
From this and from the solvability of (7.7) and (7.8) we conclude by Theorem 1.1' that $H_l(\tilde{b})$ is regular. Moreover, the condition $\varepsilon \neq 0$ implies that the matrices $H_{ml}(\tilde{b})$ and $H_{ln}(\tilde{b})$ have maximal rank and, therefore, the conditions (7.2) of Lemma 7.2 are satisfied. Hence, by Lemma 7.2,

$$B := I^o_{nl} H_l(\tilde{b})^{-1} I^o_{lm}$$

is a generalized inverse of H. Taking into account Theorem 1.1' or (1.48) we obtain (7.6), and our theorem is proved. ≡

REMARK 7.5. It is easily seen that the Theorem 7.4 also holds true if the condition $r < \min\{m,n\}$, which is equivalent to $\varkappa_1 \varkappa_2 < 0$, is replaced by the weaker condition $\varkappa_1 \varkappa_2 \leq 0$ [2)]. Thus the case of a regular matrix H is also covered. Furthermore, let us note that the algorithms described in Subsection 6.1 and 6.2 can be applied for generalized inversion of T- and H-matrices.
We show now that the converse direction of the statement of Theorem 7.4 is also true. Denote by $\mathcal{B}_{nm}(l)$ the set of all matrices B of the form

1) For the definition of I^k_{nl} see 0.2.

2) Here $\varkappa_1$, $\varkappa_2$ denote the partial indices of H.

$$B = I^o_{nl}\ C\ I^o_{lm},$$

where C is a regular classical $l\times l$ Bezoutian.

THEOREM 7.5. Let $H = H_{mn}(b)$, $b = (b_i)_0^{m+n-2}$, be an H-matrix with $r = \operatorname{rank} H < \min\{m,n\}$ and $l = m+n-r$. If B is a generalized inverse of H belonging to the class $\mathcal{B}_{nm}(l)$ then there are characteristic polynomials $p(\lambda)$ and $q(\lambda)$ ($q(\lambda)$ monic) of H such that

$$B = I^o_{nl}\ \operatorname{Bez}(q,p)\ I^o_{lm}.$$

PROOF. Suppose $B = I^o_{nl}\ C\ I^o_{lm}$ is a generalized inverse of H, where C is a regular Bezoutian. Then, by Theorem 1.1' and (1.48)

$$C = \operatorname{Bez}(\underline{y},\underline{w}),$$

where $\underline{y} = [{y \atop 0}]$, $\underline{w} = [{w \atop -1}]$ and y, w are the solutions of the fundamental equations for the H-matrix $\tilde{H} := C^{-1}$

$$\tilde{H}y = e_{l-1},\quad \tilde{H}w = (b_{l+i})_{i=0}^{l-1} \quad (b_{2l-1} \text{ arbitrary}) \qquad (7.9)$$

with $\tilde{H} = H_l(b)$, $b = (b_i)_0^{2l-2}$. From the first of these equations we obtain

$$\ker H_{l-1,l}(b) = \operatorname{lin}\{y\}. \qquad (7.10)$$

On the other hand, since H has the partial indices $(n-r, r-m)$ the matrix $H_{l-1,r+1}(b)$ has the partial indices $(1, r-l+1)$. Hence the kernel of the latter matrix is spanned by p, where $p(\lambda)$ is the first characteristic polynomial of H (uniquely determined up to a constant factor). Comparing this with (7.10) we obtain that $y = (y_0 \ldots y_r, 0\ldots 0)$ and $y(\lambda) = p(\lambda)$.

From the second equation of (7.9) one concludes that $\underline{w}(\lambda)$ is a second characteristic polynomial of H. The polynomial $\underline{w}(\lambda)$ cannot be a linear combination of $\lambda^i y(\lambda)$ $(i = 0,1,\ldots)$ because otherwise $\operatorname{Bez}(\underline{y},\underline{w})$ cannot be regular. This completes the proof. ≡

7.4. The Moore-Penrose inverse. Unfortunately the Moore-Penrose inverse of an H-matrix is, in general, no Bezoutian and no restricted Bezoutian. One way for computing the Moore-Penrose inverse bases upon the following elementary fact:

Let A be a matrix, Π_1 the orthogonal projection onto $\ker A$, Π_2 the orthogonal projections onto $\operatorname{im} A$ and B generalized inverse to A. Then

$$A^+ = (I - \Pi_1)\ B\ \Pi_2$$

is the Moore-Penrose inverse of A.

We show how the projection Π_1 and Π_2 in the case of an H-matrix $H = H_{mn}(b)$ with $r = \operatorname{rank} H < \min\{m,n\}$ can be evaluated. Let $p(\lambda)$ be a characteristic polynomial of H with lowest characteristic degree.

We introduce the $j \times k$ matrices

$$P_{jk} = [I^0 p \dots I^{k-1} p].$$

Now, the following facts are easily proved.

LEMMA 7.3. The matrices $P^*_{jk} P_{jk}$ are strongly regular T-matrices, the matrix

$$\Pi_1 := P_{n,n-r}(P^*_{n,n-r} P_{n,n-r})^{-1} P^*_{n,n-r}$$

is the orthoprojection onto $\ker H$ and

$$\Pi_2 := I_m - P_{m,m-r}(P^*_{m,m-r} P_{m,m-r})^{-1} P^*_{m,m-r}$$

the orthoprojection onto $\operatorname{im} H$.

We present now another representation of the Moore-Penrose inverse using Vandermonde matrices. For this let us introduce a notation. Let Σ be a system of the form (5.2), $\nu = \sum \nu_i$, then the generalized $n \times \nu$ Vandermonde matrix with columns $l^k_{n-1}(\lambda_i)$ $(k = 0,\dots,\nu_i-1)$ is denoted by $V_n(\Sigma)$.

PROPOSITION 7.1. Let H be an $m \times n$ H-matrix with $r = \operatorname{rank} H < \min\{m,n\}$ and let $p(\lambda) \in \mathbb{C}^{r+1}(\lambda)$ be a characteristic polynomial of H. Further, let $\Sigma = \{(\lambda_i,\nu_i)\}$ denote the system of roots λ_i of $\bar{p}(\lambda)$ with the corresponding multiplicity ν_i. Then $V_m(\Sigma)^* H V_n(\Sigma)$ is regular and

$$H^+ := V_n(\Sigma)(V_m(\Sigma)^* H V_n(\Sigma))^{-1} V_m(\Sigma)^*$$

is the Moore-Penrose inverse of H.

PROOF. Obviously, $\operatorname{im} V_m(\Sigma)$ is the orthogonal complement of $\ker H$ and $\ker V_m(\Sigma)^*$ the orthogonal complement of $\operatorname{im} H$. From these facts it is easily deduced that H^+ is just Moore-Penrose inverse to H. ≡

7.5. One-side inverses. In the previous subsections we mainly considered the case of an H-matrix, which has no maximal rank. In this subsection we shall present a possibility of generalized inversion for the maximal rank case. Clearly, in this case by generalized inversion is meant one-side inversion.

Consider an H-matrix $H := H_{mn}(b)$, $b = (b_i)_0^{m+n-2}$, and the corresponding fundamental equations

$$Hy = e_{m-1}, \quad Hw = h_m, \tag{7.11}$$

where $h_k := (b_{m+n-k+i})_{i=0}^{k-1}$ and b_{m+n-1} is arbitrary.

THEOREM 7.6. The equations (7.11) are solvable if and only if $\operatorname{rank} H = m$. In the latter case the matrix B with the columns

$$(U_n - yh_n^T + we_{n-1}^T)^{m-1-k} y \quad (k = 0,\dots,m-1)$$

is right inverse to H.

PROOF. We have obviously

$$HU_n - U_m^T H = e_{m-1} h_n^T - h_m e_{n-1}^T.$$

Supposing that the equations (7.11) are solvable we conclude

$$H(U_n - yh_n^T + w e_{n-1}^T) = U_m^T H$$

and

$$H(U_n - yh_n^T + w e_{n-1}^T)^k y = (U_m^T)^k Hy = e_{m-1-k}.$$

From this equality follows that B is right inverse to H.
On the other side, it is trivial that $\operatorname{rank} H = m$ implies the solvability of (7.11). ≡

8. CANONICAL REPRESENTATION

8.1. Definitions. In this section we deal with the problem of representation of H- and T-matrices as a linear combination of simplest H- and T-matrices. By "simplest" we understand matrices generated by vectors of the form

$$l_{pq}^k(\lambda) := (k!\binom{i}{k} \lambda^{i-k})_{i=p}^q \quad (p,q \in \mathbb{Z},\ \lambda \in \mathbb{C} - \{0\})$$

$$l_{pq}^k(0) := e_{p+k},\quad l_{pq}^k(\infty) := e_{q-k}$$

(see also Subsection 0.3). Let us explain the problem more in detail. We begin with the (simpler) case of H-matrices. The H-matrices

$$Hl^k(\lambda) := H_{mn}(l_{o,m+n-1}^k(\lambda))\ ^{1)} \quad (\lambda \in \overline{\mathbb{C}})$$

will be called elementary H-matrices. For example, we have for $\lambda \in \mathbb{C}$ $Hl^o(\lambda) = [\lambda^{i+k}]$. Obviously, $\operatorname{rank} Hl^k(\lambda) = k$.

DEFINITION 8.1. A representation of an H-matrix $H \in \mathcal{H}(m,n)$ in the form

$$H = \sum_{i=o}^{l} \sum_{k=o}^{\nu_i-1} \alpha_{ik}\, Hl^k(\lambda_i), \tag{8.1}$$

where $\lambda_i \in \overline{\mathbb{C}}$, $\alpha_{ik} \in \mathbb{C}$, $\alpha_{i,\nu_i-1} \neq 0$, will be referred to as a canonical representation of H. The number $\nu = \sum \nu_i$ is called the rank of the representation (8.1). We say by (8.1) a simple canonical representation is given if $\nu_i = 1$ and $\lambda_i \neq \infty$ for all i.
A simple canonical representation (8.1) can be written in the form

$$H = V_m(\lambda)\, \operatorname{diag}(\alpha_{1o},\ldots,\alpha_{lo})\, V_n(\lambda)^T,$$

1) In the following we shall omit the subindices of $l_{pq}^k(\lambda)$ or write for short $l_{oq}^k(\lambda) =: l_q^k(\lambda)$.

where $\lambda = (\lambda_1,\ldots,\lambda_l)$ and $V_m(\lambda)$ denotes the Vandermonde matrix $[\lambda_k^i]$ (c.f. proof of Theorem 2.6).

We are interested in canonical representations of minimal rank. It is easily observed that a canonical representation of an H-matrix $H_{mn}(b)$ automatically implies canonical representations (of the same rank) for the whole family of matrices H^k generated by b, $H^k := H_{m+n-k,k}(b)$, $k = 1,\ldots,m+n-1$. In particular, the representation (8.1) is equivalent to

$$b = \sum_i \sum_{k=0}^{\nu_i-1} \alpha_{ik}\, 1_{m+n-2}^k(\lambda_i).$$

The problem of canonical representation is, therefore, equivalent to the following generalized moment problem: For given numbers $b_0,\ldots,b_{m+n-2}$ find out a distribution with finite support on the Riemann sphere $\overline{\mathbb{C}}$

$$w = \sum_{i,k} \alpha_{ik}\, \delta^{(k)}(\lambda - \lambda_i)$$

such that

$$b_i = \int_{\overline{\mathbb{C}}} \lambda^i\, w\, d\lambda.$$

Here $\delta^{(k)}$ denotes the kth derivative of the Dirac distribution and $\delta^{(k)}(\lambda - \infty)$ will be understood as

$$\delta_{i,m+n-2-k} = \int_{\overline{\mathbb{C}}} \lambda^i\, \delta^k(\lambda - \infty) d\lambda.$$

In this language a simple canonical representation is given by

$$b_i = \int_{\mathbb{C}} \lambda^i\, d\mu,$$

where μ is a measure on $\mathbb{C}$ with finite support.

Another equivalent to the representation (8.1) will be obtained if we go over to the bilinear form

$$H(x,y) = y^*Hx \quad (x \in \mathbb{C}^n,\ y \in \mathbb{C}^m)$$

corresponding to $H \in \mathcal{H}(m,n)$.

PROPOSITION 8.1. Suppose $H \in \mathcal{H}(m,n)$ is an H-matrix with a canonical representation (8.1). Then

$$H(x,y) = \sum_i \sum_{k=0}^{\nu_i-1} \frac{\alpha_{ik}}{k!} (x(\lambda)\overline{y}(\lambda))^{(k)}(\lambda_i), \tag{8.2}$$

where $(.)^{(k)}(\lambda_i)$ denotes the k th derivative at the point $\lambda_i \in \mathbb{C}$ and

$$(x(\lambda)y(\lambda))^{(k)}(\infty) := k! \sum_{j=0}^{k} x_{n-j-1}\, y_{j+n-k-1}.$$

This proposition immediately follows from the following

LEMMA 8.1.

$$\mathrm{Hl}^k(\lambda_o)(x,y) = \frac{1}{k!}\,(x(\lambda)\bar{y}(\lambda))^{(k)}(\lambda_o). \tag{8.3}$$

PROOF. For $\lambda_o = \infty$ relation (8.3) is trivial.
Suppose now $\lambda_o \neq \infty$, then

$$\mathrm{Hl}^k(\lambda_o)(x,y) = \sum_{t=o}^{m}\sum_{s=o}^{n} \lambda_o^{s+t-k}\binom{s+t}{k} x_s y_t.$$

With the help of the identity

$$\binom{s+t}{k} = \sum_{j=o}^{k}\binom{s}{j}\binom{t}{k-j}$$

we conclude

$$\mathrm{Hl}^k(\lambda_o)(x,y) = \sum_{s,t}\sum_{j=o}^{k}\binom{s}{j}\lambda_o^{s-j}x_s\binom{t}{k-j}\lambda_o^{t-(k-j)}y_t$$

$$= \sum_{j=o}^{k}\frac{1}{j!}\,x^{(j)}(\lambda_o)\,\frac{1}{(k-j)!}\,y^{(k-j)}(\lambda_o)$$

$$= \frac{1}{k!}\sum_{j=o}^{k}\binom{k}{j}\,x^{(j)}(\lambda_o)\,y^{(k-j)}(\lambda_o).$$

This implies (8.3). ≡

Now let us consider T-matrices. The matrices

$$\mathrm{Tl}^k(\lambda) = T_{mn}^{\mu}(1_{pq}^{k}(\lambda)) \text{ [1]} \qquad (\lambda \in \overline{\mathbb{C}}),$$

where $\mu = [\frac{n-m}{2}]$ [2], $p = -m+\mu+1$, $q = n+\mu-1$, will be called <u>elementary T-matrices.</u>

<u>DEFINITION 8.2.</u> A representation of a T-matrix $T \in \mathcal{T}(m,n)$ in the form

$$T = \sum_{i=o}^{l}\sum_{k=o}^{\nu_i-1}\alpha_{ik}\,\mathrm{Tl}^k(\lambda_i), \tag{8.4}$$

where $\lambda_i \in \overline{\mathbb{C}}$, $\alpha_{ik} \in \mathbb{C}$, $\alpha_{i,\nu_i-1} \neq 0$ will be referred to as a <u>canonical representation of T.</u> The number $\nu = \sum \nu_i$ is called the <u>rank of the representation (8.4).</u> We say (8.4) is a <u>simple canonical representation</u> if $\nu_i = 1$ and $\lambda_i \neq 0,\infty$ for all i.

Analogously to the case of H-matrices the canonical representation (8.4) implies a canonical representation for the whole family of matrices $T^k := T_{m+k,n-k}^{-k}(a)$ $(k = 0,1,\dots,n-1)$, especially for the vector a. Moreover, similarly to Prop. 8.1 the following interpretation of canonical representation in the language of bilinear forms $T(x,y)=y^*Tx$ is possible.

---

[1] Such matrices were introduced in 0.2.

[2] [.] denotes the entire-function.

PROPOSITION 8.1'. Suppose $T \in \mathcal{T}(m,n)$ is a T-matrix with canonical representation (8.4). Then

$$T(x,y) = \sum_i \sum_{k=0}^{\nu_i-1} \frac{\alpha_{ik}}{k!} (x(\lambda^{-1})\overline{y}(\lambda))^{(k)}(\lambda_i),$$

where

$$(x(\lambda^{-1})\overline{y}(\lambda))^{(k)}(\infty) := k! \sum_{j=0}^{k} x_{n-k+j-1} \overline{y}_j,$$

$$(x(\lambda^{-1})\overline{y}(\lambda))^{(k)}(0) := k! \sum_{j=0}^{k} x_j \overline{y}_{n-k+j-1}.$$

8.2. The main theorem.

THEOREM 8.1. Suppose A is an m×n H- or T-matrix, n_1 and n_2 ($n_1 \leq n_2$) are its characteristic degrees. Then the following assertions hold true:

(1) A has no canonical representation of rank $\nu < n_1$.

(2) If $n_1 < n_2$ then A has a uniquely determined canonical representation (8.1) of rank $\nu < n_2$. Moreover, this representation has exactly the rank $\nu = n_1$, and

$$\prod_{\lambda_i \neq \infty} (\lambda - \lambda_i)^{\nu_i} \in \mathbb{C}^{n_1+1}(\lambda)$$

is a characteristic polynomial of A.

(3) A has infinitely many simple canonical representations with rank $\nu = n_2$. Further, if by (8.1) a canonical representation of rank n_2 is given then

$$\prod_{\lambda_i \neq \infty} (\lambda - \lambda_i)^{\nu_i} \in \mathbb{C}^{n_2+1}(\lambda)$$

is a characteristic polynomial of A.

Before proving the theorem let us note some corollaries.

COROLLARY 8.1. If A has no maximal rank then A has a unique canonical representation of rank $\nu < \max\{m,n\}$, and for this representation holds true

$$\operatorname{rank} A = \nu.$$

COROLLARY 8.2. A regular n×n H- or T-matrix has no canonical representation of rank $\nu < n$ but infinitely many simple canonical representations of rank $\nu = n$.

In the proof of Theorem 8.1 we shall make use of the following elementary observation.

REMARK 8.1. Suppose $b \in \mathbb{C}^{m+n-1}$, then the equation

$$H_{mn}(b)\, x = y \quad (x \in \mathbb{C}^n,\ y \in \mathbb{C}^m)$$

is equivalent to

$$D_m(x)\, b = y.$$

PROOF OF THEOREM 8.1. Let us prove the theorem for the Hankel case. First we show that an H-matrix H admits a canonical representation of rank n_1 if $n_1 < n_2$. Suppose $p(\lambda)$ is a characteristic polynomial of H corresponding to the characteristic degree n_1 and

$$p(\lambda) = \prod_{i=1}^{l} (\lambda - \lambda_i)^{\nu_i}, \quad \lambda_0 := \infty, \quad \nu_0 = n_1 - \sum_{i=1}^{l} \nu_i.$$

Then, by definition, p belongs to ker H^{n_1+1}, which is, in virtue of Remark 8.1, equivalent to

$$D_{n_2-1}(p)\, b = 0.\ ^{1)}$$

Applying Prop. 0.4. we obtain that b is a linear combination of the vectors $1^k(\lambda_i)$ $(i = 0,\dots,l;\ k = 0,\dots,\nu_i-1)$. Hence H is a linear combination of the matrices $H1^k(\lambda_i)$, which is a canonical representation of rank n_1.

Now we prove the uniqueness of the canonical representation of rank $\nu < n_2$. Suppose the canonical representation constructed above is given by (8.1) and

$$H = \sum_{i=0}^{l'} \sum_{k=0}^{\nu_i'-1} \alpha_{ik}'\, H1^k(\lambda_i'),$$

where $\nu = \sum \nu_i' < n_2$, $\lambda_0' = \infty$ is another canonical representation. We put

$$s(\lambda) = \prod_{i=0}^{l'} (\lambda - \lambda_i)^{\nu_i'} = \sum_k s_k \lambda^k$$

and $s = (s_k)_0^{\nu} \in \mathbb{C}^{\nu+1}$. Then by Prop. 0.4 and Remark 8.1 we have

$$H^{\nu+1} s = 0.$$

According to the definition of the characteristic polynomials we conclude that $\nu \geq n_1$ holds and s is a linear combination of the U-chain generated by p or, which is the same, $s(\lambda) = \alpha(\lambda)p(\lambda)$ for some polynomial $\alpha(\lambda)$. Therefore, the set $\{\lambda_i\}$ is a subset of $\{\lambda_i'\}$. Numbering $\lambda_0 = \lambda_0' = \infty$, $\lambda_1 = \lambda_1', \dots, \lambda_l = \lambda_l'$ we have, moreover, $\nu_i \leq \nu_i'$ $(i = 0,\dots,l)$. Comparing the two canonical representations we obtain $\nu_i = \nu_i'$, $\alpha_{ik} = \alpha_{ik}'$ for $i = 0,\dots,l$ and $\alpha_{ik}' = 0$ for

---

1) c.f. Table 5.1: $n_1 + n_2 = m + n$.

$i = l+1,\dots,l'$, because the vectors $1^k(\lambda_i')$ $(i = 0,\dots,l';\ k=0,\dots,\nu_i'-1)$ form a linearly independent system. In that manner the uniqueness is proved.

It remains to prove the third assertion. Let $q(\lambda)$ be a characteristic polynomial with characteristic degree n_2. Then for all $\varepsilon \in \mathbb{C}$ $q_\varepsilon(\lambda) := q(\lambda) + \varepsilon\,\lambda^{n_2-n_1}p(\lambda)$ is also a characteristic polynomial. By Prop. 5.5, $p(\lambda)$ and $q(\lambda)$ have no common root. Therefore, ∞ is no root of $q_\varepsilon(\lambda)$ except one value of ε. We show that $q_\varepsilon(\lambda)$ has simple roots only for almost all ε. Let λ_o be a multiple root of $q_\varepsilon(\lambda)$. Then $q(\lambda_o) + \varepsilon\tilde{p}(\lambda_o) = 0$ and $q'(\lambda_o) + \varepsilon\,\tilde{p}'(\lambda_o) = 0$, where $\tilde{p}(\lambda) := \lambda^{n_2-n_1}\,p(\lambda)$. We obtain $\tilde{p}(\lambda_o) \neq 0$, $\varepsilon = -\dfrac{q(\lambda_o)}{p(\lambda_o)}$ and $\left(\dfrac{q(\lambda)}{\tilde{p}(\lambda)}\right)'(\lambda_o) = 0$. Since the set f roots of $\left(\dfrac{q(\lambda)}{\tilde{p}(\lambda)}\right)'$ is finite, we obtain that the multiple roots of $q_\varepsilon(\lambda)$ only occur for a finite set of ε.

Now we take an ε such that $q_\varepsilon(\lambda)$ has only simple roots $\lambda_1,\dots,\lambda_{n_2}$. By definition q_ε belongs to $\ker H^{n_2+1}$, which is, in virtue of Remark 8.1, equivalent to

$$D_{n_2-1}(q_\varepsilon)\,b = 0.$$

It remains to apply Prop. 0.4, and we obtain a simple canonical representation of rank n_2.

Finally, let us assume that (8.1) is a canonical representation of rank n_2. We set $s \in \mathbb{C}^{n_2+1}$, $s(\lambda) = \prod\limits_{i=1}^{l}(\lambda - \lambda_i)^{\nu_i}$ $(\lambda_o = \infty)$. Then we have, according to Prop. 0.4,

$$D_{n_2-1}(s)\,b = 0,$$

which is, by Remark 8.1, equivalent to

$$H^{n_2+1}\,s = 0.$$

This implies that $s(\lambda)$ is a characteristic polynomial or $s(\lambda) = \alpha(\lambda)p(\lambda)$, where $n_1 > n_2$ and $\alpha(\lambda)$ is a polynomial of degree $n_2 - n_1$, and $p(\lambda)$ is a characteristic polynomial corresponding to n_1. Using the arguments of the proof of uniqueness above we conclude that the latter case is not possible. In that manner the theorem is completely proved. ≡

In the following we shall assume that in the canonical representation (8.1) of an H-matrix there is $\lambda_o = \infty$ and in the canonical representation (8.4) of a T-matrix there is $\lambda_o = 0$, $\lambda_1 = \infty$. An immediate consequence of Theorem 8.1 are the following assertions.

COROLLARY 8.3. Suppose (8.1) is a canonical representation of rank n_1, then $(\sum\limits_{i=1}^{l}\nu_i, \nu_o)$ is just the (ϱ,π)-characteristic of the H-matrix H. In particular, H is quasi-regular if and only if H admits a canonical representation (8.1) of rank n_1 with $\lambda_i \neq \infty$ for all i.

COROLLARY 8.3'. Suppose (8.4) is a canonical representation of rank n_1, then $(\sum_{i=1}^{l-1} \nu_i, \nu_0, \nu_l)$ is just the (ϱ, π_+, π_-)-characteristic of T. In particular, T is half-regular if and only if T admits a canonical representation (8.4) of rank n_1 with $\lambda_i \neq 0(\infty)$ for all i.

8.3. Canonical representation of Hermitian H- and T-matrices.

In this subsection we discuss the specific of Hermitian H- and T-matrices. An H-matrix is Hermitian, clearly, if and only if it is square and real. Real H-matrices have real characteristic polynomials [1]. Therefore an Hermitian H-matrix has a canonical representation (8.1) such that the numbers λ_i are real or pairwise conjugate complex. Moreover, one observes, as a consequence of the linear independence of the vectors $l^k(\lambda_i)$, that the numbers α_{ik} corresponding to conjugate complex λ_i are conjugate complex, and they are real if λ_i are real. Respecting these observations we obtain the following theorem, which we shall formulate in the language of quadratic forms (c.f. Prop. 8.1).

THEOREM 8.2. Let $H(x,x) = x^*Hx$, $H \in \mathcal{H}(n)$ be an Hermitian H-form. Then there is a representation of the form

$$\begin{aligned} H(x,x) = &\sum_{i=1}^{l} \sum_{k=0}^{\nu_i-1} \frac{1}{k!} \operatorname{Re}\beta_{ik} \, (x(\lambda)\overline{x}(\lambda))^{(k)}(\lambda_i) + \\ &+ \sum_{j=0}^{m} \sum_{k=0}^{\varrho_j-1} \frac{1}{k!} \gamma_{jk} \, (x(\lambda)\overline{x}(\lambda))^{(k)}(\varepsilon_j), \end{aligned} \tag{8.5}$$

where

$$2 \sum \nu_i + \sum \varrho_j = \operatorname{rank} H,$$

$\beta_{ik} \in \mathbb{C}$, $\gamma_{jk} \in \mathbb{R}$, λ_i are non-real, $\varepsilon_0 = \infty$ and ε_j $(j > 0)$ are real numbers. This representation is unique if and only if H is non-regular. Moreover, the polynomial with the roots λ_i, $\overline{\lambda}_i$ and ε_j is a characteristic polynomial of H. [2]

Our aim is now to represent an H-form as a sum of squares using only the characteristic polynomial $p(\lambda)$.

LEMMA 8.2. The following identities hold true

$$(1)\ \operatorname{Re} \sum_{k=0}^{\nu-1} \frac{\beta_k}{k!}(x(\lambda)\overline{x}(\lambda))^{(k)} = \frac{1}{4} \sum_{k=0}^{\nu-1} \Big[\Big| \sum_{s=k}^{\nu-1} \frac{\beta_s}{s!}\binom{s}{k} x^{(k-s)}(\lambda) + \overline{x}^{(s)}(\lambda) \Big|^2 - \Big| \sum_{s=k}^{\nu-1} \frac{\beta_s}{s!} \binom{s}{k} x^{(k-s)}(\lambda) - \overline{x}^{(s)}(\lambda) \Big|^2 \Big],$$

[1] Let us note that the converse is also true, i.e. if the characteristic polynomial $p(\lambda)$ is real then H is real.

[2] It can be formulated, according to Theorem 8.1, a stronger (but also longer) version of this theorem.

$$(2)\ \sum_{k=0}^{2\sigma-1} \gamma_k (x(\varepsilon)\bar{x}(\varepsilon))^{(k)} = 2\ \mathrm{Re} \sum_{s=0}^{\sigma-1} [\tfrac{1}{2}\binom{2s}{s}\ \gamma_{2s}\ x^{(s)}(\varepsilon) + \sum_{t=s+1}^{2\sigma-1-s} \binom{s+t}{s}\ \gamma_{s+t}\ x^{(t)}(\varepsilon)]\ \bar{x}^{(s)}(\varepsilon),$$

$$(3)\ \sum_{k=0}^{2\sigma} \gamma_k (x(\varepsilon)\bar{x}(\varepsilon))^{(k)} = \binom{2\sigma}{\sigma}\ \gamma_{2\sigma} |x^{(\sigma)}(\varepsilon)|^2 + \sum_{k=0}^{2\sigma-1} \gamma_k (x(\varepsilon)\bar{x}(\varepsilon))^{(k)}.$$

Here $\beta_k \in \mathbb{C}$, $\lambda \in \mathbb{C}$, $\gamma_k \in \mathbb{R}$, $\varepsilon \in \mathbb{R} \cup \{\infty\}$, $x(\lambda) \in \mathbb{C}^n(\lambda)$.

The proof is an elementary computation. From Lemma 8.2 and Theorem 8.2 we conclude

THEOREM 8.3. Let $H(x,x) = x^*Hx$ be an Hermitian H-form and

$$p(\lambda) = \prod_i (\lambda - \lambda_i)^{\nu_i} (\lambda - \bar{\lambda}_i)^{\nu_i} \prod_j (\lambda - \varepsilon_j)^{\varrho_j} \in \mathbb{C}^r(\lambda),$$

where $\lambda_i \notin \mathbb{R}$, $\varepsilon_j \in \mathbb{R}$, $r = \mathrm{rank}\ H$, be a characteristic polynomial of H. Then this form can be represented as a sum of independent squares as follows

$$H(x,x) = \sum_i \sum_{k=0}^{\nu_i-1} (|X_{ik}^+|^2 - |X_{ik}^-|^2) + \sum_j \sum_{k=0}^{\sigma_j-1} (|Z_{ik}^+|^2 - |Z_{ik}^-|^2) + \sum_{\varrho_j \text{ odd}} \binom{\varrho_j-1}{\sigma_j} \gamma_{\varrho_j-1} |x^{(\sigma_j)}(\varepsilon)|^2,$$

where $\sigma_j = [\tfrac{1}{2}\varrho_j]$,

$$X_{ik}^{\pm} = \sum_{s=k}^{\nu_i-1} \frac{\alpha_{is}}{s!} \binom{s}{k} x^{(k-s)}(\lambda_i) \pm \bar{x}^{(s)}(\lambda_i)$$

and

$$Z_{ik}^{\pm} = \tfrac{1}{2}\binom{2k}{k}\gamma_{j,2k} x^{(k)}(\varepsilon_i) + \sum_{t=k+1}^{2\sigma_j-1-k} \binom{k+t}{k}\gamma_{j,k+t} x^{(2\sigma_j-1-k)}(\varepsilon_i) \pm \bar{x}^{(s)}(\varepsilon_i)$$

COROLLARY 8.4. Let $H(x,x)$ be an Hermitian H-form with a canonical representation (8.5) of minimal rank. Then

$$\mathrm{sgn}\ H = \sum_{\varrho_j \text{ odd}} \mathrm{sgn}\ \gamma_{j,\varrho_j-1}.$$

In particular $H \in \mathscr{H}(n)$ is positive definite if and only if H has a characteristic polynomial with n simple real roots.

From this corollary one can easily deduce the Jacobi-Borchardt theorem (Theorem 2.6), Prop. 2.16 and others.
Let us consider now Hermitian T-matrices $T \in \mathscr{T}(n)$. From Prop. 0.2 we conclude that if $p(\lambda)$ $(q(\lambda))$ is a characteristic polynomial with

characteristic degree n_1 (n_2) then $\overline{p}(\lambda^{-1})\lambda^{n_1}$ $(\overline{q}(\lambda^{-1})\lambda^{n_2})$ is also a characteristic polynomial. Therefore, there exist symmetric characteristic polynomials $p(\lambda) \in \mathbb{C}^{n_1+1}$, $q(\lambda) \in \mathbb{C}^{n_2+1}(\lambda)$ of T and the roots of the symmetric polynomial[1] $p(\lambda)$ are symmetrically situated relative to the unit circle. Let ζ_j $(j = 1,\dots,m)$ be the roots of $p(\lambda)$ with $|\zeta_j| = 1$ and ϱ_j their multiplicities. Further, let λ_i $(i = 0,1,\dots,l)$ be the roots of $p(\lambda)$ with $|\lambda_i| < 1$ and ν_i the corresponding multiplicities. As a modification of Prop. 0.4 one can show that the vectors

$$l_j^k := l_{1-n,n-1}^k(\zeta_j) \quad (j = 1,\dots,m;\ k = 0,\dots,\varrho_j-1)$$

$$l_i^{k,+} := l_{1-n,n-1}^k(\lambda_i);\ l_i^{k,-} := J_{2n-1}\overline{l}_i^{k,+} \quad (i=0,\dots,l;\ k=0,\dots,\nu_i-1)$$

form a basis of $\ker D_n(p)$ (c.f. Subsection 0.3).

<u>THEOREM 8.4.</u> Let $T(x,x) = x^*Tx$ be an Hermitian T-form. Then there is a representation of $T(x,x)$ of the form

$$T(x,x) = \sum_{i=0}^{l} \sum_{k=0}^{\nu_i-1} \frac{1}{k!} \operatorname{Re} \beta_{ik} \, (x(\lambda^{-1})\overline{x}(\lambda))^{(k)}(\lambda_i) + {}$$

$$+ \sum_{j=1}^{m} \sum_{k=0}^{\varrho_j-1} \frac{1}{k!} \gamma_{jk} \, (x(\lambda^{-1})\overline{x}(\lambda))^{(k)}(\zeta_j),$$

where

$$2 \sum \nu_i + \sum \varrho_j = \operatorname{rank} T,$$

$\beta_{ik} \in \mathbb{C}$, $\gamma_{ik} \in \mathbb{T}$, $|\lambda_i| < 1$, $|\zeta_j| = 1$. This representation is unique if and only if T is non-regular. Moreover, the polynomial with the roots λ_i, $\overline{\lambda}_i^{-1}$ and ζ_j is a characteristic polynomial of T.

<u>8.4. Canonical representation of singular extensions.</u> In this subsection let $Hl_{mn}^k(\lambda)$ denote the mxn elementary H-matrix $Hl^k(\lambda)$ and $Tl_{mn}^k(\lambda)$ the mxn elementary T-matrix $Tl^k(\lambda)$. Obviously $Hl_{m+\mu,n+\nu}^k(\lambda)$ $(\mu,\nu = 0,1,\dots)$ is a singular (μ,ν)-extension of $Hl_{mn}^k(\lambda)$ [2] if and only if $\lambda \neq \infty$, $Tl_{m+\mu,n+\nu}^k(\lambda)$ is a half-singular extension of $Tl_{mn}^k(\lambda)$ for all $\lambda \in \mathbb{T}$, which is singular if and only if $\lambda \neq 0,\infty$. This observation together with Theorem 8.1 leads to the following propositions.

<u>PROPOSITION 8.2.</u> Suppose $H \in \mathcal{H}(m,n)$ admits a singular (μ,ν)-extension and

$$H = \sum_i \sum_{k=0}^{\nu_i-1} \alpha_{ik} \, Hl_{mn}^k(\lambda_i) \tag{8.6}$$

1) c.f. Subsection 4.5.
2) c.f. Subsection 5.8.

is a canonical representation with $\sum \nu_i$ = rank H. Then

$$\tilde{H} = \sum_i \sum_{k=0}^{\nu_i-1} \alpha_{ik}\, Hl^k_{m+\mu,n+\nu}(\lambda_i) \tag{8.7}$$

is a singular (μ,ν)-extension of H. Conversely, if (8.7) is the canonical representation of a (μ,ν)-extension of H with $\sum \nu_i$ = rank H, then (8.6) is a canonical representation of H.

PROPOSITION 8.3. Suppose $T \in \mathcal{T}(m,n)$ is (+)-half-regular ((−)-half-regular) and admits a half-regular (μ,ν)-extension. Let further

$$T = \sum_{i=0}^{l} \sum_{k=0}^{\nu_i-1} \alpha_{ik}\, Tl^k_{mn}(\lambda_i) \tag{8.8}$$

be a canonical representation of T, where $\sum \nu_i$ = rank H, $\lambda_o = 0$ or $\lambda_o = \infty$. Then

$$\tilde{T} = \sum_{i=1}^{l} \sum_{k=0}^{\nu_i-1} \alpha_{ik}\, Tl^k_{m+\mu,n+\nu}(\lambda_i) + \sum_{k=0}^{\nu_o-1} \alpha_{ok}\, Tl^{k+\delta}_{m+\mu,n+\nu}(\lambda_o) \tag{8.9}$$

with

$$\delta := \begin{cases} \mu & \text{if } \lambda_o = 0 \\ \nu & \text{if } \lambda_o = \infty \end{cases} \tag{8.10}$$

is a half-singular (μ,ν)-extension of T. Conversely, if (8.10) is the canonical representation of a half-singular extension of T with $\sum \nu_i$ = rank T then (8.8) is a canonical representation of T.

8.5. Symbol description of strongly singular H- and T-matrices.

We proceed now the description of the classes of strongly singular H- and T-matrices. As we know (c.f. Prop. 5.14) strongly singular H-matrices are singular (∞,∞)-extensions of quasi-regular H-matrices, and strongly singular T-matrices are half-singular (∞,∞)-extensions of half-regular T-matrices. Introducing the infinite elementary H- and T-matrices

$$Hl^k_\infty = [k!\,\tbinom{i+j}{k}\,\lambda^{i+j}]^\infty_{i,j=0} \quad (\lambda \neq 0),$$

$$Tl^k_\infty = [k!\,\tbinom{i-j}{k}\,\lambda^{i-j}]^\infty_0 \quad (\lambda \neq 0,\infty)$$

we obtain a canonical representation of strongly singular H- and T-matrices.

PROPOSITION 8.4. A strongly singular H-matrix H admits a unique representation in the form

$$H = \sum_i \sum_{k=0}^{\nu_i-1} \alpha_{ik}\, Hl^k_\infty(\lambda_i). \tag{8.11}$$

A strongly singular T-matrix T admits a unique representation in the form

$$T = \sum_i \sum_{k=0}^{\nu_i - 1} \alpha_{ik} \, Tl_\infty^k(\lambda_i) + S, \tag{8.12}$$

where S is a T-matrix, $S = [s_{i-j}]_0^\infty$, such that $s_i = 0$ for all $i \leq \nu$ or for all $i \geq -\nu$ with $\nu := \sum \nu_i$.

The first assertion immediately follows from Prop. 8.2 and 5.14. The second assertion yields taking into account that (8.12) is equivalent to

$$T_n = \sum_i \sum_{k=0}^{\nu_i - 1} \alpha_{ik} \, Tl^k(\lambda_i) + S_n,$$

where T_n denotes the nxn principal sections of T and

$$S_n := \sum_{k=0}^{n-1-\nu} s_{n-1-k} \, Tl^k(0) \quad \text{or} \quad S_n := \sum_{k=0}^{n-1-\nu} s_{1-n+k} \, Tl^k(\infty).$$

Prop. 8.4 can be formulated in a more convenient form if one uses the concept of the symbol of an infinite H- or T-matrix.

<u>DEFINITION 8.3.</u>
(1) Suppose $b = (b_k)_0^\infty$ is such that the series

$$b(z) = \sum_{k=0}^{\infty} b_k \, z^{-k-1}$$

converges for some z, then the function $b(z)$ (defined in a neighbourhood of infinity) is referred to as the <u>symbol of the H-matrix $H_\infty(b)$.</u>

(2) Suppose $a = (a_k)_{-\infty}^\infty$ is such that

$$a(z) = \sum_{k=-\infty}^{\infty} a_k \, z^k$$

converges for almost all $z \in \mathbb{T}$ [1] then $a(z)$ is referred to as the <u>symbol of the T-matrix $T_\infty(a)$.</u>

The symbol of an elementary H-matrix is easily computed. Indeed, the matrix $Hl_\infty^k(\lambda)$ has the symbol

$$\sum_{j=0}^{\infty} \binom{j}{k} \lambda^{j-k} z^{-j-1} = \frac{d^k}{d\lambda^k} \frac{1}{z-\lambda} = \frac{k!}{(z-\lambda)^{k+1}} .$$

This observation and Prop. 8.4 yield the following statement, which is a stronger version of a well-known theorem due to Kronecker.

<u>THEOREM 8.5.</u> An H-matrix is strongly singular if and only if it has a symbol, which is a rational function.

---

[1] The convergence can be also understood in another sense.

The Toeplitz case is something more complicated, because elementary T-matrices do not have a symbol.
Let $M_-(\mathbb{T})$ denote the space of functions of the form $\varphi_-(z) = \sum_{k=-\infty}^{-1} \varphi_k z^k$ $(z \in \mathbb{T})$, where the convergence is almost everywhere.

LEMMA 8.3. Let T be a strongly singular T-matrix with the canonical representation (8.12), where $s_i = 0$ for $i \geq -\nu$ and $a(z)$ the symbol of T. Then

$$a(z) = \sum_i \sum_{k=0}^{\nu_i - 1} \alpha_{ik} \frac{\lambda_i^{-k-\nu}}{k!} z^k \frac{d^k}{dz^k} \frac{z^{-\nu}}{1-\lambda_i z} + z^{-\nu} \psi_-(z), \qquad (8.13)$$

where $\psi_-(z) \in M_-(\mathbb{T})$.

PROOF. By definition we have for $|z| > \max|\lambda_i|$

$$a(z) = \sum_i \sum_{k=0}^{\nu_i - 1} \alpha_{ik} \sum_{j=-\nu}^{\infty} \binom{j}{k} \lambda_i^{j-k} z^j + z^{-\nu} \psi_-(z)$$

for some $\psi_-(z) \in M_-(\mathbb{T})$. It follows

$$a(z) = \sum_i \sum_{k=0}^{\nu_i - 1} \frac{\alpha_{ik}}{k!} z^k \lambda_i^{-k} \frac{d^k}{dz^k} \sum_{j=-\nu}^{\infty} \lambda_i^j z^j + z^{-\nu} \psi_-(z)$$

and (8.13). ≡

The function $z^k \frac{d^k}{dz^k} \frac{z^{-\nu}}{1-\lambda_i z}$ has the poles λ_i^{-1} of multiplicity $k+1$ and 0 of multiplicity ν. Therefore, the linear hull spanned by this functions for $k = 0,1,\ldots,m$ coincides with the linear hull of $z^{-\nu}(1 - \lambda_i z)^{-k-1}$. Hence a function $a(z)$ admits a representation (8.13) if and only if it admits a representation of the form

$$a(z) = z^{-\nu}\Big(\sum_i \sum_{k=0}^{\nu_i - 1} \frac{\beta_{ik}}{(1-\lambda_i z)^{k+1}} + \varphi_-(z)\Big) \qquad \varphi_- \in M_-(\mathbb{T}).$$

In that manner the following assertion is proved.

THEOREM 8.6. A T-matrix with the symbol $a(z)$ is strongly singular if and only if $a(z)$ or $a(z^{-1})$ can be represented in the form $z^{-\nu}(r(z) + \varphi_-(z))$, where $\varphi_-(z) \in M_-(\mathbb{T})$ and $r(z)$ is a rational function with not more than ν poles (counting multiplicities).

NOTES AND COMMENTS

Section 1.

1) The first inversion formula for (finite) Toeplitz matrices occurring in the literature is the formula of our second inversion variant (Theorem 1.2). It was proved by G. BAXTER, I.I.HIRSCHMAN [1] under some needless additional assumptions using corresponding results for infinite Toeplitz matrices (well-known from the Wiener-Hopf theory). A pure algebraic proof without these assumptions was given by I. GOHBERG, A. SEMENCUL [1] (see also I. GOHBERG, I. FELDMAN [1]). The formula of the third inversion variant was proved by I. GOHBERG, N. KRUPNIK [1]. The formula of our first inversion variant was founded independently by E.M. RUSSAKOVSKI [2] and G. HEINIG [7,8] (see also G. HEINIG, K. ROST [2]). All proofs of the inversion formulae presented here are simpler than the original ones.

2) The first inversion formula applicable to arbitrary regular T-matrices was founded by A.L. SAHNOVICH [1]. Because this formula is rather complicated, we omitted it in our book. We remark that in G. HEINIG, K. ROST [2,3] it is shown how this formula can be obtained using the approach of Section 1.

3) All inversion formulae can be generalized without any essential difficulty to block T- and H-matrices. But in the block case the inversion formulae of the type (1.15), (1.28), (1.33) contain the solutions of 4 block vector equations (see I. GOHBERG, G. HEINIG [1,2] and references there).

4) PROBLEM. Can the inverse of an arbitrary regular T-matrix be constructed by means of the solutions of two fixed equations $T_n(a)x_i = g_i$ $(i=1,2)$ (i.e., g_i do not depend on $T_n(a)$)? Our guess is that the answer will be negative.

5) Knowing the fundamental solutions of $T_n(a)$ the solution of a Toeplitz system $T_n(a)\varphi = \psi$ can be constructed using one of the inversion formulae. Computing in usual way this requires $O(n^2)$ operations, but using the Fast Fourier Technique the amount can be reduced to $O(n \log n)$ operations (see H. NUSSBAUMER [1]).

Section 2.

1) The fact that the inverses of H- and T-matrices are Bezoutians (and vice versa) was first explicitly formulated by I.F. LANDER [1]. From this paper Prop. 2.15 is taken, too.

2) Concerning Barnett's formula see S. BARNETT [1], Prop. 2.13 was founded independently by N.K. KRAVITSKY [1] and E.M. RUSSAKOWSKI (private communication 1981).

3) A quite complete review of classical results concerning root localization problems is presented in the paper M. KREIN, M. NEIMARK [1] (see also F.R. GANTMACHER [1], M.M. POSTNIKOV [1] and others).

4) The concepts of resultant matrix and Bezoutians as well as some of the corresponding results can be generalized to matrix and operator polynomials. It is remarkable that two different but both natural generalizations are possible. The first generalization ("one-side" or "matrix" resultants and Bezoutians) is considered in I. GOHBERG, G.HEINIG [3], G. HEINIG [5,7], B. ANDERSON, E. JURY [1], L. LERER, M. TISMENATSKY [1]; the second generalization ("mixed" or "tensor" resultants and Bezoutians) is considered in I. GOHBERG, L. LERER [1], G. HEINIG [2,3,5,7], G. HEINIG, U. JUNGNICKEL [2], S. BARNETT, P. LANCASTER [1].
5) The presented approach for solving Liapunov and Stein equations is due to P.C. MÜLLER [1], R.A. SMITH [1], R.E. HARTWIG [1,2]. Some further results are taken from G. HEINIG, U. JUNGNICKEL [1].

Section 3.
1) It is remarkable that inversion algorithms for (strongly regular) T-matrices were constructed before inversion formulae were established. The first algorithm (for positive definite T-matrices) goes back to N. LEVINSON [1]. The further development of this algorithm is presented in the papers W. TRENCH [1], S. ZOHAR [1], L.M. KUTIKOV [1] and many others.
2) The recursion formulae can be generalized to the block case without any difficulty (see P. WHITTLE [1], L.M. KUTIKOV [2], H. AKAIKE [1], J. RISSANEN [1], I. GOHBERG, G. HEINIG [2]).
3) The presented algorithm of Bezoutian inversion is closed to the Routh algorithm (see G. HEINIG, U. JUNGNICKEL [1]).
4) An algorithm of T-matrix inversion using Fast Fourier Technique is given in R. BITMEAD, B. ANDERSON [1] (see also S.L JUNG [1]).

Section 4.
1) The sufficiency part of Theorem 4.2 (assertion b,c) was already proved by E. FISCHER [1] in a particular and by G. FROBENIUS [1] in the general case (see I.S. IOHVIDOV [7,8]). The transformation was used to reduce the trigonometric moment problem to the algebraic one. The other results are taken from G. HEINIG [6,8].
2) With the help of Möbius matrices the inversion formulae of Section 1 can be generalized (see G. HEINIG [9]).
3) Concerning the Schur-Cohn theorem see Note 3.4 to Section 2. A generalization of this theorem was recently given by V. PTAK, N. YOUNG [1].

Section 5.
1) The concept of U-chain was first introduced by M. KREIN [1] in the infinite-dimensional case (in this paper the formulation "d-chain" is used). The concept of U-chain in $\mathbb{C}^n$ was introduced in G. HEINIG [1].
2) Corollary 5.3 was proved in G. HEINIG [1] using the properties of the resultant matrix. The general kernel structure theorem is published for the first time. Furthermore, the concept of characteristic poly-

nomials of an H-matrix is introduced for the first time.
3) The concepts of (ϱ,π)- and (ϱ,π_+,π_-)-characteristic and the theorems 5.7, 5.8 are due to I.S. IOHVIDOV [5,8,9]. In these papers singular extensions were investigated, too. Half-singular extension were considered in G. HEINIG [1].
4) There are some first results concerning the generalization of the results of this section to the block case (see G. HEINIG [4]), but these are far from being complete.

Section 6.
1) The algorithm of T- and H-matrix inversion was recently published in G. HEINIG [11]. The generalization to the block case is not done yet.
2) The results about the generalized LU-factorization are published for the first time. As it has been recently shown by U. JUNGNICKEL (unpublished), Theorem 6.3 can be proved directly using Euklid's algorithm.
3) Prop. 6.1 is due to G. FROBENIUS (see F.R. GANTMACHER [1]), Prop. 6.2 to I.S. IOHVIDOV [1,8]. We note that the presented formulation of the signature rules and the corresponding proofs are simpler than the original ones.

Section 7. The results are published for the first time.

Section 8.
1) In the case of positive semidefinite T- and H-matrices Theorem 8.1 follows from the classical theory of moment problems. The case of square T-matrices was considered in G. HEINIG [1], the general case (in some weaker version) in G. HEINIG [8].
2) In G. HEINIG [1] Theorem 8.6 is used to obtain a criterion of normal solvability for a class of two-dimensional Toeplitz operators.

Additional remark.

After writing this part we took notice that in the report

Bultheel, A.:
Fast factorization algorithms and Padé approximation: a survey. K.U. Leuven, Report TW 42, 1979.

factorizations of H- and T- matrices are constructed, which are similar (but not equal) to the generalized LU-factorizations of Subsection 6.3.

PART II

TOEPLITZ-LIKE OPERATORS

We are going to show in this part that some of the results presented in Part I can be extended to other classes of matrices and operators. In the infinite-dimensional case we make use of some well-known facts from the theory of Fredholm operators we shall collect in Section 0. The reader, who is familiar with the theory of Fredholm operators or only interested in the finite-dimensional case, may omit this section.

0. FREDHOLM OPERATORS

Let $\mathbb{E}$ and $\mathbb{F}$ be complex Banach spaces with norm $\|\cdot\|$, $\mathcal{L}(\mathbb{E},\mathbb{F})$ the space of linear bounded operators $A : \mathbb{E} \to \mathbb{F}$ and $\mathbb{E}' = \mathcal{L}(\mathbb{E},\mathbb{C})$. The Banach-adjoint of A will be denoted by A'. The operator A is called <u>normally solvable</u> if its range $\operatorname{im} A$ is closed. It is called <u>Φ_+-operator</u> (<u>Φ_--operator</u>) if, moreover, $\dim \ker A < \infty$ ($\dim \mathbb{F}|\operatorname{im} A < \infty$). If A is normally solvable, then, by the Hahn-Banach theorem

$$\dim \mathbb{F}|\operatorname{im} A = \dim \ker A'.$$

An operator is said to be <u>Fredholm</u> if it is both Φ_+- and Φ_--operator. The number

$$\operatorname{ind} A = \dim \ker A - \dim \mathbb{F}|\operatorname{im} A$$

is called <u>index</u> of the Φ_+- or Φ_--operator A.

<u>PROPOSITION 0.1.</u> The operator $A \in \mathcal{L}(\mathbb{E},\mathbb{F})$ is normally solvable if and only if it has the following property: The vector $y \in \mathbb{F}$ belongs to $\operatorname{im} A$ if and only if $\varphi(y) = 0$ for all $\varphi \in \ker A'$.

<u>PROPOSITION 0.2.</u> Suppose $A \in \mathcal{L}(\mathbb{E},\mathbb{F})$ is a Φ_+- or Φ_--operator and $T \in \mathcal{L}(\mathbb{E},\mathbb{F})$ is compact. Then $A + T$ is also Φ_+- or Φ_--operator, respectively, and

$$\operatorname{ind}(A + T) = \operatorname{ind} A.$$

<u>PROPOSITION 0.3.</u> Suppose $A \in \mathcal{L}(\mathbb{E},\mathbb{F})$ is a Φ_+- or Φ_--operator. Then there is an $\varepsilon > 0$ such that for all B with $\|B\| < \varepsilon$ $A + B$ is also Φ_+- or Φ_--operator, respectively, and

$$\operatorname{ind}(A + B) = \operatorname{ind} A,$$

$$\dim \ker(A + B) \leq \dim \ker A, \ \dim \ker(A + B)' \leq \dim \ker A'.$$

An operator $R \in \mathcal{L}(\mathbb{E},\mathbb{F})$ is said to be <u>a left (right) regularizator</u> of A if $I - RA$ $(I - AR)$ is compact. If R is both left and right regularizator R is refered to as regularizator.

PROPOSITION 0.4. An operator $A \in \mathcal{L}(\mathbb{E},\mathbb{F})$ has a left (right) regularizator if and only if it is a Φ_+- (Φ_--operator) and there exists a continuous projection onto im A (ker A). The operator has a regularizator if and only if it is Fredholm.

1. TOEPLITZ-LIKE OPERATORS - FIRST GENERAL CONSIDERATIONS

1.1. Definitions. Throughout this section $\mathbb{E}$ and $\mathbb{F}$ will be two complex Banach spaces with norm $\|\cdot\|$. The space of linear bounded operators $A : \mathbb{E} \to \mathbb{F}$ will be denoted by $\mathcal{L}(\mathbb{E},\mathbb{F})$. In particular, we set $\mathcal{L} := \mathcal{L}(\mathbb{E},\mathbb{F})$, $\mathcal{L}(\mathbb{E}) := \mathcal{L}(\mathbb{E},\mathbb{E})$ and $\mathbb{E}' := \mathcal{L}(\mathbb{E},\mathbb{C})$. The Banach-adjoint operator of A will be denoted by A', $(A'\varphi)(x) = \varphi(A(x))$ $(x \in \mathbb{E},\ \varphi \in \mathbb{F}')$. Mostly we shall use for vectors Latin and for functionals Greek letters. We fix for all further investigations two operators $U \in \mathcal{L}(\mathbb{E})$ and $V \in \mathcal{L}(\mathbb{F})$, which we shall call reduction operators.

DEFINITION 1.1. (1) The operator

$$D_A := AU - VA : \mathbb{E} \to \mathbb{F}$$

is called UV-reduction of $A \in \mathcal{L}$.

(2) The rank of D_A is called UV-rank of A and will be denoted by $r(A)$.

(3) An operator is said to be a UV-Toeplitz-like operator, for short TLO, if it has a finite UV-rank.

In finite-dimensional spaces, clearly, all linear operators are TLO, but in this case the forthcoming considerations will be reasonable only if the UV-rank is small comparing with the dimension of the spaces.

The classical example for TLO in finite-dimensional spaces are operators generated by Toeplitz matrices (as well as Hankel matrices). If one chooses $U = V$ as shift cperators one obtains that Toeplitz matrices have a UV-rank less or equal 2. [1)]
We note some obvious properties.

PROPOSITION 1.1.

(1) Suppose $A, B \in \mathcal{L}$, then

$$r(A + B) \leq r(A) + r(B).$$

(2) Suppose $A \in \mathcal{L}$, $B \in \mathcal{L}(\mathbb{F},\mathbb{G})$, $W \in \mathcal{L}(\mathbb{G})$. Then

$$r(BA) \leq r(A) + r(B),$$

where $r(BA)$ denotes the UW-rank, $r(A)$ the UV-rank and $r(B)$ the VW-rank.

(3) Suppose $A \in \mathcal{L}$ is invertible, then

$r(A^{-1}) = r(A)$, where $r(A^{-1})$ denotes the VU-rank of A^{-1}.

1) c.f. I, Prop. 0.3.

We introduce now some operators, which will play the role of the matrices H^k (see I, (5.4)) in the theory of Hankel matrices.

DEFINITION 1.2.

(1) For a given operator $B \in \mathcal{L}$ the maps

$$\mathrm{Obs}_k\, B := \begin{bmatrix} B \\ B\,U \\ \vdots \\ B\,U^k \end{bmatrix} : \mathbb{E} \to \mathbb{F}^{k+1}$$

will be called observability operators of B.

(2) The maps

$$\mathrm{Con}_k\, B := [B \ \ VB \ \dots \ V^kB] : \mathbb{E}^{k+1} \to \mathbb{F}$$

will be called controllability operators of B.

(3) We shall say that $A \in \mathcal{L}$ has a complete UV-reduction if and only if there is a k_0 such that for all $k \geq k_0$

$$\ker \mathrm{Obs}_k\, A = \{0\} \quad \text{and} \quad \ker \mathrm{Con}_k\, A = \{0\}. \tag{1.1}$$

It is easy to verify that the following relations hold true

$$\ker \mathrm{Obs}_k A = \ker\begin{bmatrix} A \\ \mathrm{Obs}_{k-1}\, D_A \end{bmatrix}, \quad \ker(\mathrm{Con}_k A)' = \ker[A \ \mathrm{Con}_{k-1} D_A]'. \tag{1.2}$$

Let us note that the completeness condition (1.1) will be important for obtaining of invertibility criteria.

Now we are going to discuss this condition for the case of square T-matrices. We put $U = V = U_n$, where U_n is the forward shift in $\mathbb{C}^n$. In case $T_n(a)$ is not lower triangular we have, obviously,

$$\ker \mathrm{Obs}_k\, T_n(a) = \ker T^{n+k}$$

and

$$\ker(\mathrm{Con}_k\, T_n(a))' = \ker(T^{n-k})',$$

where $T^k := T_{2n-k,k}(a)$. Hence (1.1) is satisfied. In case $T_n(a)$ is lower triangular (1.1) is satisfied only if $T_n(a)$ is regular. To cover also this "singular" case of non-regular triangular T-matrices we must extend Definition 1.1. This extension will be needed for the generalization of the U-chain concept, too.

Besides the reduction operators U and V we fix a third operator $N \in \mathcal{L}$ called normalization operator. We shall always assume that N has finite rank. We introduce the following subspaces

$$\mathcal{G}_A = \operatorname{im} D_A + \operatorname{im} N, \quad \mathcal{F}_A = \operatorname{im} D'_A + \operatorname{im} N'.$$

DEFINITION 1.3. Let A be a TLO.

(1) The number $r_N(A) = \max\{\dim \mathcal{G}_A, \dim \mathcal{F}_A\}$ will be called UVN-rank of A.

(2) The following two conditions will be called <u>completeness</u> <u>conditions</u> for the operator A

$$\ker \begin{bmatrix} \mathrm{Obs}_k A \\ \mathrm{Obs}_{k-1} N \end{bmatrix} = 0 \qquad (k \geq k_0) \quad , \tag{C1}$$

$$\mathrm{im}\,[\mathrm{Con}_k A \quad \mathrm{Con}_{k-1} N] = F \quad (k \geq k_0). \tag{C2}$$

In the case of mxn Toeplitz matrices a natural choice of N is

$$N(x_k)_0^{n-1} = x_{n-1}e_0 \in \mathbb{C}^m.$$

Choosing N in that manner the conditions (C1) and (C2) are satisfied for any T-matrix.

<u>PROPOSITION 1.2.</u> If a TLO A satisfies the conditions (C1), (C2), then it is Fredholm.

<u>PROOF.</u> The operator $[A \quad 0]$ is a finite-dimensional perturbation of

$$[A \quad \mathrm{Con}_{k-1} D_A \quad \mathrm{Con}_{k-1} N],$$

which is, by (C2) and (1.2) an epimorphism for $k \geq k_0$. Therefore, A is a Φ_--operator, according to Prop. 0.2. Since, moreover, (C1) and (1.2) imply that $\dim \ker A < \infty$, A is Fredholm. ≡

<u>1.2. Criteria of invertibility.</u> In this subsection we shall see that, in generalization of the T-matrix case, the invertibility of a TLO yields from the solvability of "fundamental" equations, provided that the completeness conditions are satisfied.

Let $\{g_1,\ldots,g_r\}$ be a basis of $\mathrm{im}\, D_A + \mathrm{im}\, N$ and $\{\gamma_1,\ldots,\gamma_{r'}\}$ a basis of $\mathrm{im}\, D_A' + \mathrm{im}\, N'$.

The equations

$$A\, x_i = g_i \qquad (i = 1,\ldots,r) \tag{1.3}$$

and

$$A'\, \xi_j = \gamma_j \qquad (j = 1,\ldots,r') \tag{1.4}$$

will be referred to as <u>fundamental</u> <u>equations</u> corresponding to A.

In the following we shall need the relation

$$\ker D_A \cap \ker N = \bigcap_{j=1}^{r'} \ker \gamma_j. \tag{1.5}$$

This relation holds true, since $D_A z = 0$ is equivalent to $\gamma(z) = 0$ for all $\gamma \in \mathrm{im}\, D_A'$ and $Nz = 0$ is equivalent to $\gamma(z) = 0$ for all $\gamma \in \mathrm{im}\, N'$.

<u>THEOREM 1.1.</u> Suppose A is a TLO such that the completeness conditions (C1), (C2) are satisfied. Then the solvability of the fundamental equations (1.3), (1.4) implies the invertibility of A.

The theorem can be divided into two propositions.

PROPOSITION 1.3. Suppose A is a TLO such that the condition (C1) is fulfilled. Then the solvability of the equations (1.4) implies $\ker A = \{0\}$.

PROOF. Let z be an element of $\ker A$. Then we have

$$\gamma_j(z) = (A' \xi_j)(z) = 0 \qquad (j = 1,\dots,r').$$

In virtue of (1.5), this implies

$$D_A z = 0 \quad \text{and} \quad Nz = 0.$$

Hence

$$AUz = D_A z = 0,$$

i.e. $Uz \in \ker A$. Repeating all arguments we obtain next $U^2z \in \ker A$ and $NUz = 0$ and, furthermore, $U^k z \in \ker A$ and $NU^{k-1}z = 0$ for $k = 3,4,\dots$. Therefore, $z \in \mathrm{Obs}_k A$ for all k. Taking into account (C1) we conclude $z = 0$, and the proposition is proved. ≡

Going over to the adjoint operator we obtain

PROPOSITION 1.3'. Suppose A is a TLO such that the condition (C2) is fulfilled. Then the solvability of the equations (1.3) implies $\operatorname{im} A = \mathbb{F}$.

In the following we shall always assume (without loss of generality) that the vectors $g_1,\dots,g_r$ form a basis of $\operatorname{im} D_A$. Then the functionals $\gamma_1,\dots,\gamma_r$ ($\in \operatorname{im} D_A'$) can be chosen such that

$$D_A = \sum_{i=1}^{r} \gamma_i(\cdot)\, g_i. \tag{1.6}$$

We explain now how the invertibility criteria from the second and third inversion variant of T-matrices (c.f. I, Subsections 1.2, 1.3) can be generalized. Let $h_1,\dots,h_p$ be eigenvectors of V corresponding to the eigenvalues $\lambda_1,\dots,\lambda_p$. Consider instead of (1.3) the following system of equations

$$\begin{cases} A\, y_i = h_i & (i = 1,\dots,p) \\ A\, x_i = g_i & (i = p+1,\dots,r), \end{cases} \tag{1.7}$$

where $p \le r$.

PROPOSITION 1.4. Suppose the TLO $A \in \mathcal{L}$ satisfies the condition (C2), the equations (1.7) are solvable and the $p \times p$ matrix

$$P := [\gamma_i(y_k)]_1^p$$

is regular. Then $\operatorname{im} A = \mathbb{F}$.

PROOF. Since P is assumed to be regular, the system with unknown vectors $x_1,\dots,x_p$

$$\sum_{k=1}^{p} \gamma_k(y_i)x_k = (U-\lambda_i I)y_i - \sum_{k=p+1}^{r} \gamma_k(y_i)x_k \quad (i=1,\ldots,p) \tag{1.8}$$

has a unique solution $\{x_k : k = 1,\ldots,p\}$. We show that $A\, x_k = g_k$ holds. Indeed, we have

$$A(U - \lambda_i I)y_i = A(U - \lambda_i I)y_i - (V - \lambda_i I)Ay_i = D_A y_i$$
$$= \sum_{k=p+1}^{r} \gamma_k(y_i)g_k + \sum_{k=1}^{p} \gamma_k(y_i)Ax_k.$$

This implies, in virtue of (1.8),

$$\sum_{k=1}^{p} \gamma_k(y_i)Ax_k = \sum_{k=1}^{p} \gamma_k(y_i)g_k.$$

Using the regularity of P we conclude $A\, x_k = g_k$ for $k = 1,\ldots,p$. It remains to apply now Prop. 1.3', and the assertion is proved. ≡

REMARK 1.1. Assuming the conditions of Prop. 1.4 are satisfied, the solutions of (1.3) for $i = 1,\ldots,p$ can be expressed by the solutions of (1.6) via the formula

$$x_i = \sum_{k=1}^{p} c_{ik}[(U - \lambda_k I)y_k - \sum_{s=p+1}^{r} \gamma_s(y_k)x_s], \tag{1.9}$$

where $[c_{ik}] = P^{-1}$.

The formula (1.9) immediately follows from (1.8).

An analogous proposition holds if the system (1.4) is replaced by a system

$$\begin{cases} A'\ \chi_i = \omega_i & (i = 1,\ldots,q) \\ A'\ \xi_i = \gamma_i & (i = q+1,\ldots,r'), \end{cases} \tag{1.10}$$

where ω_i are eigenvectors of U', provided that the matrix

$$Q := [\chi_i(g_k)]_1^q$$

is regular.

Now we present a generalization of the invertibility criterion for the third inversion variant of T-matrices. We consider instead of the system (1.3) a system

$$A\, x_{ik} = V^k g_i \quad (k = 0,\ldots,r_i-1,\ i = p+1,\ldots,r), \tag{1.11}$$

where $p \le r$, $\sum_{i=p+1}^{r} r_i = r$ and $r_i > 1$ for $i = p+1,\ldots,q$.

PROPOSITION 1.5. Suppose the TLO $A \in \mathcal{L}$ satisfies the condition (C2), the equations (1.11) are solvable, and the pxp matrix

$$\tilde{P} := [\gamma_j(x_{i,k-1})]_{\substack{j=1,\ldots,p \\ k=1,\ldots,r_i-1;\, i=p+1,\ldots,q}}$$

is regular. Then $\operatorname{im} A = \mathbb{F}$.

PROOF. Consider the following system of equations with unknown vectors $x_1,\dots,x_p$

$$\sum_{j=1}^{p} \gamma_j(x_{i,k-1})x_j = U\, x_{i,k-1} - x_{ik} - \sum_{j=p+1}^{r}\gamma_j(x_{i,k-1})x_j \quad (1.12)$$

$$(k = 1,\dots,r_i-1;\ i = p+1,\dots,q).$$

Since the matrix of coefficients $\mathfrak{P}$ is assumed to be regular, this system has a unique solution $\{x_1,\dots,x_p\}$. We show that $A\, x_j = g_j$ holds. Indeed, we have

$$A(Ux_{i,k-1}-x_{ik}) = VA\, x_{i,k-1} - A\, x_{ik} + D_A\, x_{i,k-1} = D_A\, x_{i,k-1}.$$

Hence, (1.12) implies

$$\sum_{j=1}^{p} \gamma_j(x_{i,k-1})\, A\, x_j = \sum_{j=1}^{p} \gamma_j(x_{i,k-1})g_j.$$

Taking into account the regularity of $\mathfrak{P}$ we conclude $A\, x_j = g_j$. It remains to apply Prop. 1.3'. ≡

REMARK 1.2. Assuming the conditions of Prop. 1.5 are satisfied, the solutions of (1.3) for $i = 1,\dots,p$ can be expressed by the solutions of (1.11) via the formula

$$x_j = \sum_{i,k} c_{jik}(U\, x_{i,k-1} - x_{ik} - \sum_{s=p+1}^{r}\gamma_s(x_{i,k-1})x_s),$$

where

$$\mathfrak{P}^{-1} = [c_{jik}]_{\substack{j=1,\dots,p\\ k=1,\dots,r_i-1;\, i=p+1,\dots,q}} .$$

1.3. Solution of special equations.

Let $A \in \mathcal{L}$ be a TLO and

$$D_A = \sum_{i=1}^{r} \gamma_i(\cdot)g_i.$$

Throughout this subsection we shall assume that the fundamental equations

$$A\, x_i = g_i \qquad (i = 1,\dots,r) \tag{1.13a}$$

are solvable.[1] Moreover, we assume that for a fixed vector $e \in \mathbb{F}$ the equation

$$A\, x = e \tag{1.13b}$$

has a solution.

For convenience we introduce operators $G : \mathbb{C}^r \to \mathbb{F}$, $X : \mathbb{C}^r \to \mathbb{E}$, $\Gamma : \mathbb{E} \to \mathbb{C}^r$ by

[1] We do not assume here that the completeness conditions are fulfilled.

$$G(a_i)_1^r := \sum_{i=1}^{r} a_i g_i \qquad (a_i \in \mathbb{C})$$

$$X(a_i)_1^r := \sum_{i=1}^{r} a_i x_i$$

$$\Gamma u := (\gamma_i(u))_1^r \qquad (u \in \mathbb{E}).$$

Then we have

$$D_A = G\Gamma \quad \text{and} \quad A(U - X\Gamma) = VA. \tag{1.14}$$

We introduce now the concept of transfer function of an operator triple, which is originated in linear system theory.

DEFINITION 1.4. Suppose $\Phi : \mathbb{E} \to \mathbb{C}^n$, $C : \mathbb{E} \to \mathbb{E}$ and $K : \mathbb{C}^n \to \mathbb{E}$ are linear operators. Then the nxn matrix function

$$\theta(\mu) := \theta(\Phi,C,K)(\mu) := I_n - \mu \Phi (I - \mu C)^{-1} K$$

is called transfer function of the triple (Φ,C,K). $\theta(\mu)$ is defined for all $\mu \in \mathbb{C}$ for which $(I - \mu C)^{-1}$ exists. This set will be denoted by $\varrho^{-1}(C)$ and referred to as inverse resolvent set of C.

PROPOSITION 1.6. Suppose the equations (1.13) are solvable. Then the equation

$$A\, x(\mu) = (I - \mu V)^{-1} e \tag{1.15}$$

is solvable for all $\mu \in \varrho^{-1}(V) \cap \varrho^{-1}(U - X\Gamma)$, and its solution is given by the formula

$$x(\mu) = (I + \mu(I - \mu U)^{-1}\, \theta(\mu)^{-1}\Gamma)(I - \mu U)^{-1} x, \tag{1.16}$$

where $\theta(\mu) = \theta(\Gamma,U,X)(\mu)$.

PROOF. Suppose $\mu \in \varrho^{-1}(V) \cap \varrho^{-1}(U - X\Gamma)$, then by (1.14) we have

$$A(I - \mu(U - X\Gamma))^{-1} = (I - \mu V)^{-1} A.$$

From this relation we conclude that (1.15) is solvable and

$$x(\mu) = (I - \mu(U - X\Gamma))^{-1} x$$

is a solution of (1.15). It can be verified by straightforward calculation that $\theta(\mu)$ is regular and

$$(I - \mu(U - X\Gamma))^{-1} = (I - \mu U)^{-1} + \mu(I-\mu U)^{-1}\theta(\mu)^{-1}\, \Gamma(I - \mu U)^{-1}.$$

This yields (1.16) and completes the proof. ≡

In practice it is often not easy to compute the solution $x(\mu)$ via formula (1.16). We are going to present a more convenient rule for computing this solution. To obtain such a rule we have to assume that the equations

$$A' \xi_i = \gamma_i \qquad (i = 1,\dots,r) \tag{1.17}$$

are solvable, where γ_i are given by (1.6).

PROPOSITION 1.7. Suppose the equations (1.13), (1.17) are solvable. Then (1.15) is solvable for all $\mu \in \varrho^{-1}(U) \cap \varrho^{-1}(V - G\Xi)$, and a solution is given by

$$x(\mu) = (I - \mu U)^{-1}(x - \mu X \Xi (I - \mu V)^{-1} e), \tag{1.18}$$

where $\Xi := \mathbb{F} \to \mathbb{C}^r$ is defined by

$$\Xi y = (\xi_i(y))_1^r.$$

PROOF. We have $AU = (V - G\Xi)A$, hence

$$A(I - \mu U)^{-1} = (I - \mu(V - G\Xi))^{-1} A. \tag{1.19}$$

Let $x(\mu)$ be defined by (1.18). Then we obtain respecting the latter relation

$$A\,x(\mu) = (I - \mu(V - G\Xi))^{-1} A(x - \mu X \Xi (I - \mu V)^{-1} e)$$

$$= (I - \mu V)^{-1} e,$$

and the proposition is proved. ≡

REMARK 1.3. The relation (1.18) holds for all $\mu \in \varrho^{-1}(U)$ if A is an invertible operator (and in some other cases).

This remark immediately follows from equality (1.19).

In the remainder of this subsection we discuss (heuristically) some possibilities of the construction of inversion formulae. Is is a matter of fact that in many particular cases the formulae for the special solutions (1.15) already leads to formulae for right-side inverse or inverse operators.

Consider, for example, the case $\mathbb{F} = \mathbb{C}^m$, $V = U_n$ (the forward shift) and $e = e_o$. Then we have $(I - \mu V)^{-1} e = \sum_{k=o}^{m-1} \mu^k e_k$, where e_k is the (k+1)th unit vector. Supposing

$$x(\mu) = x_o + x_1\mu + \dots$$

we obtain after comparing the coefficients in (1.15)

$$A\,x_k = e_k.$$

Therefore, the matrix with the columns x_k $(k = 0,\dots,m-1)$ is a right inverse of A.

This approach can be also applied if $\mathbb{F}$ is a Banach space of sequences $(x_k)_0^\infty$ $(x_k \in \mathbb{C})$ and V is the shift operator in $\mathbb{F}$.
We have still another possibility of the construction of a right inverse operator in case that any vector $y \in \mathbb{F}$ can be represented in the form

$$y = \int (I - \mu V)^{-1} e \, dm_y,$$

where m_y is a measure on $\mathbb{C}$ depending on y. Indeed, in that case it is reasonable to expect that

$$u = \int x(\mu) \, dm_y \tag{1.20}$$

is a solution of $Au = y$ and, therefore, the operator $y \to \int x(\mu) \, dm_y$ is a right-inverse of A. This idea can be applied in the special case of $\mathbb{E} = \mathbb{F} = L_p[0,\tau]$ $(\tau \in \mathbb{R})$ if V is the integration operator and $e = e(t) \equiv 1$. In this case we have, as is easily verified,

$$(I - \mu V)^{-1} e = e^{\mu t}$$

and, therefore, a formula of the form (1.20) can be obtained using Fourier technique.

2. INVERSION FORMULAE FOR TOEPLITZ-LIKE MATRICES

This and the next sections are concerned with the finite-dimensional case $\mathbb{E} = \mathbb{C}^n$, $\mathbb{F} = \mathbb{C}^m$, i.e. we consider Toeplitz-like operators $A : \mathbb{C}^n \to \mathbb{C}^m$. As in Part I we shall always identify such operators with the corresponding matrix. In this sense we use the concept of a Toeplitz-like matrix. In what follows we make use of all notations and abbreviations introduced in Section 0 of Part I.
The main goal of this section is to generalize the inversion formulae for T- and H-matrices of I, Section 1 to Toeplitz-like matrices using the approach outlined in Section 1. We specify our investigation to the following three classes of UV-Toeplitz-like matrices :

1) matrices close to Toeplitz - by choosing U and V as shifts (analogously, matrices close to Hankel can be considered),
2) matrices close to Vandermonde - by choosing U as diagonal matrix and V as shift,
3) matrices of the form $[(c_i - b_k)^{-1}]$ and their generalizations - by choosing U and V both as diagonal matrices.

Furthermore, we consider matrices of the form Toeplitz plus Hankel more in detail.

<u>2.1. Matrices close to Toeplitz.</u> As above we denote by U_n the forward shift in $\mathbb{C}^n$, $U_n := [\delta_{i-1,k}]$. In this subsection we put for the

reduction operators U, V and the normalization operator N

$$U := U_n, \quad V := U_m, \quad N := e_o\, e_m^T.$$

This choice of N guarantees the fulfilment of the completeness conditions (C1) and (C2) for an arbitrary matrix. U_nU_m-Toeplitz-like matrices, i.e. m×n matrices A, for which rank $(AU_n - U_mA)$ is small comparing with m and n, will be called <u>close to Toeplitz</u>, because the T-matrices are the simplest representatives of this class. The number

$$\dim(\mathrm{im}(AU_n - U_mA) + \mathrm{lin}\{e_o\})$$

will be referred to as <u>Toeplitz-rank</u> (for short <u>T-rank</u>) of A.
Let us present some examples of matrices close to Toeplitz.

1) T-Bezoutians (see I, Section 2), in particular inverses of T-matrices, have a T-rank ≤ 3. In generalization of this fact, inverses of matrices close to Toeplitz are close to Toeplitz again. Moreover, the T-ranks of A and A^{-1} differ at most by ± 1.
2) A sum of products of a few T-matrices is close to Toeplitz. For example, let T_1, T_2, T_3 be T-matrices of suitable size, then $T_1T_2 + T_3$ has T-rank ≤ 6. Another important example are (Hermitian) matrices of the form T^*T, where T is Toeplitz, which have a T-rank ≤ 4.

We shall see later that any matrix A with T-rank r can be represented in the form $A = \sum_{i=o}^{r} T_i^+ T_i^-$, where T_i^+ are lower and T_i^- upper triangular T-matrices.

3) Matrices of the form

$$A = [T_1\ T_2], \tag{2.1}$$

where T_1, T_2 are Toeplitz are called <u>paired T-matrices.</u> Such matrices occur when solving paired discrete Wiener-Hopf equations as well as singular integral equations by the reduction method.
We evaluate D_A for such matrices. Suppose

$$T_1 = T_{ml}(a) \quad (a = (a_i)_{-l}^{m-1}), \quad T_2 = T_{m,n-l}(b) \quad (b = (b_i)_{l-n}^{m-1})$$

then

$$D_A = g_1\, e_{l-1}^T + g_2\, e_{n-1}^T + e_o\, f^T,$$

where

$$g_1 = (b_o - a_{-l} \ldots b_{m-1} - a_{m-l-1}), \quad g_2 = -(b_{l-n} \ldots b_{m+l-n-1}),$$
$$f = (a_{-1} \ldots a_{-l},\ \ b_{-1} \ldots b_{l-n}). \tag{2.2}$$

Hence A has a T-rank ≤ 3. A special type of paired T-matrices are resultant matrices (see I, Subsection 2.5).
In generalization of paired T-matrices one can consider matrices

of the form

$$A = \begin{bmatrix} T_{11} & \cdots & T_{1t} \\ \vdots & & \vdots \\ T_{s1} & \cdots & T_{st} \end{bmatrix}, \tag{2.3}$$

where T_{ij} are T-matrices. Matrices of this form will be referred to as <u>matrices of T-mosaic type.</u> It is easily verified that a matrix (2.3) has a T-rank $\leq t+s$.

<u>REMARK 2.1.</u> An equation of the form

$$Ax = y, \tag{2.4}$$

where A is a sum of products of $n \times n$ T-matrices, is equivalent to a system with a coefficient matrix of the form (2.3). More in detail, suppose $A = \sum_{i=1}^{r} T_i^+ T_i^-$, where $T_i^{\pm}$ is Toeplitz, then (2.4) is equivalent to

$$\mathfrak{X}\,\tilde{\mathfrak{x}} = \tilde{\mathfrak{y}}, \tag{2.5}$$

where

$$\mathfrak{X} := \begin{bmatrix} 0 & T_1^+ & \cdots & T_r^+ \\ T_1^- & -I_n & & \\ \vdots & & \ddots & \\ T_r^- & & & -I_n \end{bmatrix}, \quad \tilde{\mathfrak{y}} := \begin{bmatrix} y \\ 0 \\ \vdots \\ 0 \end{bmatrix}, \quad \tilde{\mathfrak{x}} := \begin{bmatrix} x \\ x_1 \\ \vdots \\ x_r \end{bmatrix}.$$

4) Block-Toeplitz matrices $[A_{i-k}]_{0}^{m-1}{}_{0}^{n-1}$ with $d \times d$ blocks A_i can be transformed changing the order of rows and columns into the matrices of T-mosaic type (2.3), where $s = t = d$.
5) The consideration of $n \times n$ matrices of the form $A = T + H$, where T is Toeplitz and H is Hankel, can be replaced (under some assumption) by the consideration of the matrix

   $$\begin{bmatrix} T & HJ_n \\ J_nH & J_nTJ_n \end{bmatrix},$$

   which is a matrix of T-mosaic type with $s = t = 2$. We shall explain this more in detail in Section 2.2.

We are going to construct inversion formulae for matrices closed to Toeplitz, which are generalizations of the corresponding formulae for T-matrices (c.f. I, Section 1). Let A be an $m \times n$ matrix with T-rank $r+1$. Then $D_A = AU_n - U_mA$ can be represented in the form

$$D_A = \sum_{i=0}^{r} g_i\, f_i^T. \tag{2.6}$$

We shall always assume that $g_o = e_o$. In case e_o does not belong to im D_A we put $f_0 := 0$. The equations

$$Ax_i = g_i \qquad (i = 0,\dots,r) \tag{2.7}$$

and

$$A^T y_i = f_i \qquad (i = 0,\dots,r) \tag{2.8}$$

will be referred to as <u>fundamental</u> <u>equations</u> corresponding to A. These equations can also be written in matrix form

$$AX = G, \tag{2.7'}$$

$$A^T Y = F, \tag{2.8'}$$

where G, F, X, Y denote the matrices with the columns g_i, f_i, x_i, y_i, respectively.

<u>THEOREM 2.1.</u> Suppose the equations (2.7) are solvable. Then rank $A = m$, and the matrix C with the columns c_k $(k = 0,\dots,m-1)$,

$$c_0 = x_0, \quad c_{k+1} = (U_n - XF^T)c_k \tag{2.9}$$

is a right inverse of A.

<u>PROOF.</u> By Prop. 1.3', we have $\operatorname{im} A = \mathbb{C}^m$, which is equivalent to rank $A = m$. From (2.6) and (2.7') we obtain (compare with (1.14))

$$A(U_n - XF^T) = U_n A.$$

Hence,

$$A(U_n - XF^T)^k\, x_0 = U_n^k\, Ax_0 = e_k,$$

and we conclude $A\, c_k = e_k$, which completes the proof. ≡

<u>THEOREM 2.2.</u> Suppose A is a regular $n \times n$ matrix and (2.6) holds. Then the inverse of A is given by the formula

$$A^{-1} = T_n(x_0) - \sum_{i=0}^{r} T_n(x_i)\, T_n(U_n y_i)^T, \tag{2.10}$$

where x_i, y_i are the solutions of (2.7), (2.8), respectively.

<u>PROOF.</u> According to I, Lemma 1.1 we have

$$A^{-1}\, U_n^k = U_n^k\, A^{-1} - \sum_{s=0}^{k-1} U_n^s\, X\, Y^T\, U_n^{k-1-s}\ .$$

Hence, for the k th column c_k $(k = 0,\dots,n-1)$ of A^{-1} the following relation holds true

$$c_k = U_n^k\, x_0 - \sum_{s=0}^{k-1} U_n^s\, X\, Y^T\, e_{k-1-s},$$

which is another expression of (2.10). ≡

REMARK 2.2. The columns c_k of A^{-1} can also be evaluated via the following recursion

$$c_o = x_o, \quad c_{k+1} = U_n c_k - X Y^T e_k \quad (k = 0,\dots,n-2). \tag{2.11}$$

This recursion is a generalization of formula I, (1.18) for T-matrices.

REMARK 2.3. The computation of A^{-1} using the recursion (2.9) requires $2(r+1)n(n-1)$ multiplications, it requires $(r+1)n(n-1)$ multiplications if one uses the recursion (2.11).

Now we show how to establish the matrix A knowing D_A and the first column of A. The result will also make clear that any matrix closed to Toeplitz can be represented as the sum of a few products of triangular T-matrices.

THEOREM 2.3. Suppose A is an mxn matrix $(m \le n)$, and a is the first column of A. Then (2.6) holds if and only if

$$A = T_{mn}(a) + \sum_{i=o}^{r} T_{mn}(g_i)\, T_n(U_n f_i)^T \tag{2.12}$$

and

$$\sum_{i=o}^{r} T_{mn}(g_i)\, \hat{f}_i = 0, \tag{2.13}$$

where $\hat{f}_i := J_n f_i$ with $J_n[\delta_{i,n-j-1}]_o^{n-1}$.

PROOF. Suppose (2.6) holds. Then, by means of I, Lemma 1.1, we obtain (2.12). Furthermore, (2.6) implies

$$0 = AU_n e_{n-1} = U_n A e_{n-1} + G F^T e_{n-1},$$

which is another expression of (2.13).

Conversely, let A be given by (2.12). Then we have

$$D_A = \sum_{i=o}^{r} T_{mn}(g_i)(T_n(U_n f_i)^T U_n - U_n T_n(U_n f_i)^T)$$

$$= \sum_{i=o}^{r} g_i f_i^T - (\sum_{i=o}^{r} T_{mn}(g_i)\, \hat{f}_i) e_{n-1}^T.$$

Therefore (2.6) holds, provided that (2.13) is satisfied. $\equiv$

As for T-matrices, the formulae (2.10) and (2.12) can be expressed in terms of generating functions. For this we have to evaluate ΔA^{-1} (respectively ΔA) introduced in I, Subsection 1.6.

Let C be an arbitrary mxn matrix. Then ΔC has the following structure

$$\Delta C = \begin{bmatrix} c_o & D_C \\ 0 & -c'_{m-1} \end{bmatrix},$$

where c_o is the first column and c'_{m-1} the last row of C. Hence

$$(\Delta C)(\lambda,\mu) = c_o(\lambda) - \lambda^{m-1}\mu c'_{m-1}(\mu) + \mu\, D_C(\lambda,\mu).$$

Assuming $D_C = \sum_{i=o}^{r} u_i v_i^T$ we conclude

$$C(\lambda,\mu) = \frac{1}{1-\lambda\mu}(c_o(\lambda) - \lambda^{m-1}\mu c'_{m-1}(\mu) + \mu \sum_{i=o}^{r} u_i(\lambda)\, v_i(\mu)\,).$$

This formula leads to expressions for the generating functions of A^{-1} and A.

In the remainder of this subsection we shall discuss the specific of some special cases of matrices close to Toeplitz. Let us consider first matrices of T-mosaic type. In this case it seems to be natural to modify the reduction operators. More precisely speaking, we replace $U = U_n$, $V = U_m$ by the following block-diagonal matrices

$$U = \operatorname{diag}(U_{n_1}\cdots U_{n_t}),\quad V = \operatorname{diag}(U_{m_1}\cdots U_{m_s}), \tag{2.14}$$

where $m_i \times n_k$ is the size of the block T_{ik}. In the special case of paired T-matrices (2.1) this means that $U = \operatorname{diag}(U_l, U_{n-l})$, $V = U_m$ and, consequently,

$$AU - VA = g_1\, e_{l-1}^T + g_2\, e_{n-1}^T + e_o\, f^T,$$

where $g_1 = -(a_{-l}\cdots a_{m-l-1})$ and g_2, f are defined as in (2.2). We obtain, in particular, the following assertion.

<u>PROPOSITION 2.1.</u> Suppose A is a regular paired T-matrix. Then A^{-1} can be represented in the form

$$A^{-1} = \begin{bmatrix} T_{ln}(x_o) \\ T_{n-l,n}(x'_o) \end{bmatrix} - \sum_{i=o}^{2} \begin{bmatrix} T_{ln}(x_i) \\ T_{n-l,n}(x'_i) \end{bmatrix} T_n(U_n y_i)^T,$$

where $x'_i := I_{n-l,n}^{-l+1}\, x_i \quad (i = 0,1,2).$[1)]

Choosing the reduction matrices U and V according to (2.14) one may apply the second inversion variant (see Prop. 1.4). In case $s = t = 2$ we obviously have $e_{l-1}, e_{n-1} \in \ker V$ and $e_o, e_l \in \ker U^T$, where $l \times l'$ is the size of T_{11}. This means that A^{-1} can be constructed with the help of the solutions of the following equations

$$Au_k = e_k \quad (k = 0,l-1,l,n-1),\quad Av_j = e_j \quad (j = 0,l'-1,l',n-1).$$

Clearly, these solutions are columns or rows of the inverse matrix

1) The definition of I_{mn}^k c.f. I, Subsection 0.2.

$A^{-1} = [c_{ik}]$. But, according to Prop. 1.4, this requires the fulfilment of the additional condition that the two matrices

$$P := \begin{bmatrix} c_{l-1,l-1} & c_{l-1,n-1} \\ c_{n-1,l-1} & c_{n-1,n-1} \end{bmatrix}, \qquad Q := \begin{bmatrix} c_{oo} & c_{ol'} \\ c_{l',o} & c_{l',l'} \end{bmatrix} \tag{2.15}$$

are regular.

Finally we discuss the specific of a symmetric matrix A closed to Toeplitz. It may be expected that in this case a formula of the form (2.10) can be established using only the solutions of r+4 equations. On the other hand, since D_A is not symmetric, there is no direct way to obtain such a formula. We have to replace D_A by a symmetric modification of this matrix. Suppose $A = [a_{ik}]_o^{n-1}$ and define

$$A_1 := [a_{ik}]_o^{n-2}, \quad A_2 := [a_{ik}]_1^{n-1}, \quad \bar{D}_A := A_2 - A_1.$$

Then $\bar{D}_A$ is symmetric and

$$D_A = \begin{bmatrix} a^T & 0 \\ \bar{D}_A & -b \end{bmatrix}, \tag{2.16}$$

where $a = (a_{oi})_1^{n-1}$, $b = (a_{i,n-1})_o^{n-2}$. The matrix D_A^T can be represented in the form

$$\bar{D}_A = \sum_{i=1}^{r} h_i \, h_i^T = H \, H^T \qquad (h_i \in \mathbb{C}^{n-1}), \tag{2.17}$$

where h_i are assumed to be linearly independent. In virtue of (2.16) and Theorem 2.2, the inverse of A can be constructed by means of the solutions of

$$A \, x_i = \begin{bmatrix} h_i \\ 0 \end{bmatrix}, \quad A \, y_i = \begin{bmatrix} 0 \\ h_i \end{bmatrix} \qquad (i = 1,\dots,r), \tag{2.18}$$

$$A \, x_o = e_o, \qquad A \, y_o = e_{n-1}, \qquad A \, x_{r+1} = \begin{bmatrix} 0 \\ b \end{bmatrix}, \quad A \, y_{r+1} = \begin{bmatrix} a \\ 0 \end{bmatrix}.$$

Provided that A_1, A_2 are regular, it is easily seen that the solutions of (2.18) can be constructed with the help of the solutions of

$$A_1 \, x_i' = h_i, \quad A_2 \, y_i' = h_i \qquad (i = 1,\dots,r) \tag{2.19}$$

via the formulae

$$x_i = \begin{bmatrix} x_i' \\ 0 \end{bmatrix} - \alpha_i \, y_o, \qquad y_i = \begin{bmatrix} 0 \\ y_i' \end{bmatrix} - \beta_i \, x_o, \tag{2.20}$$

where $\alpha_i = b^T x_i'$, $\beta_i = a^T y_i'$. Let X', Y' be the matrices with columns x_i', y_i', respectively. Then (2.17), (2.19) imply, assuming the regularity of A_2,

$$Y'(I_r + H^T X') = X'.$$

We show that $I_r + H^T X'$ is regular. Suppose $(I_r + H^T X')\varphi = 0$, then we have

$$A_2 X'\varphi = A_1 X'\varphi + HH^T X'\varphi = H(I_r + H^T X')\varphi = 0,$$

which implies $X'\varphi = 0$ and $\varphi = 0$. We obtain

$$Y' = (I_r + H^T X')^{-1} X' \tag{2.21}$$

and have proved the following assertion.

<u>PROPOSITION 2.2.</u> Let A be a symmetric nxn matrix such that the sections A_1 and A_2 are regular. Then A^{-1} can be constructed with the help of the solutions x_i' $(i = 1,\dots,r)$, x_0, y_0, x_{r+1} and y_{r+1} via the formulae (2.21), (2.20) and (2.10).

<u>2.2. Toeplitz plus Hankel matrices.</u> This subsection is concerned with matrices of the form $A = T + H$, where T is an nxn Toeplitz and H is an nxn Hankel matrix. We present two approaches for inverting matrices of this form. The first approach bases upon the relationship between the matrix A and the 2nx2n matrix of T-mosaic type

$$\tilde{A} := \begin{bmatrix} T & HJ_n \\ J_nH & J_nTJ_n \end{bmatrix}.$$

<u>PROPOSITION 2.3.</u> The matrix $\tilde{A}$ is regular if and only if the two matrices $A = T + H$ and $A_- = T - H$ are regular. If $\tilde{A}$ is regular, then the inverse of A is given by

$$A^{-1} = [I_n \quad 0]\, \tilde{A}^{-1} \begin{bmatrix} I_n \\ J_n \end{bmatrix}. \tag{2.22}$$

<u>PROOF.</u> Let $C_{\pm}$ denote the subspaces of $\mathbb{C}^{2n}$ consisting of all vectors of the form $(x, \pm J_n x)$ $(x \in \mathbb{C}^n)$. Obviously $\mathbb{C}^{2n}$ is the direct sum of C_+ and C_-. Let us investigate the acting of $\tilde{A}$ on $C_{\pm}$. We have

$$\tilde{A} \begin{bmatrix} x \\ J_nx \end{bmatrix} = \begin{bmatrix} Ax \\ J_nAx \end{bmatrix} \text{ and } \tilde{A} \begin{bmatrix} x \\ -J_nx \end{bmatrix} = \begin{bmatrix} A_-x \\ -J_nA_-x \end{bmatrix}. \tag{2.23}$$

That means the subspaces C_+ and C_- are invariant under $\tilde{A}$. Furthermore, we conclude from the latter relations that the restriction of $\tilde{A}$ onto C_+ is similar to A, and the restriction onto C_- is similar to A_-. Therefore, $\tilde{A}$ is regular if and only if A and A_- are regular. Finally, the first of the relations (2.23) immediately yields (2.22). ≡

Now we discuss the inversion of the T-mosaic type matrix $\mathfrak{X}$. We denote as usual $T = T_n(a)$, $a = (a_k)_{-n}^{n}$, $H = H_n(b)$, $b = (b_i)_{-1}^{2n-1}$ and take, as reduction operators, $U = V = \operatorname{diag}(U_n, U_n)$. Then we have

$$\mathfrak{X}U - U\mathfrak{X} = \sum_{i=1}^{4} g_i f_i^T,$$

where $g_1 = e_0$, $g_2 = e_n$, $f_3 = e_{n-1}$, $f_4 = e_{2n-1}$,

$g_3 = -(a_{-n}\dots a_{-1}, b_{2n-1}\dots b_{n-2})$, $g_4 = -(b_{-1}\dots b_{n-2}, a_n\dots a_1)$,

$f_1 = (a_{-1}\dots a_{-n}, b_{n-2}\dots b_{-1})$, $f_2 = (b_n\dots b_{2n-1}, a_1\dots a_n)$.

The inverse of $\mathfrak{X}$ can be constructed with the help of the solutions of

$$\mathfrak{X}\, x_i = g_i \tag{2.24a}$$

and $\qquad (i = 1,\dots,4)$

$$\mathfrak{X}^T y_i = f_i \tag{2.24b}$$

Let us exploit the special structure of $\mathfrak{X}$. We have, obviously,

$$J_{2n}\, \mathfrak{X}\, J_{2n} = \mathfrak{X} \tag{2.25a}$$

and

$$\begin{bmatrix} J_n & 0 \\ 0 & J_n \end{bmatrix} \mathfrak{X} \begin{bmatrix} J_n & 0 \\ 0 & J_n \end{bmatrix} = \mathfrak{X}^T. \tag{2.25b}$$

The relation (2.25b) implies

$$y_3 = \operatorname{diag}(J_n, J_n)x_1 \quad \text{and} \quad y_4 = \operatorname{diag}(J_n, J_n)x_2.$$

In order to simplify the system (2.24) we apply the second inversion variant. Due to Subsection 2.1 the equations (2.24a) for $i = 3,4$ can be replaced by

$$\mathfrak{X}\, z_1 = e_{n-1}, \quad \mathfrak{X}\, z_2 = e_{2n-1}$$

and, analogously, (2.24b) for $i = 1,2$ by

$$\mathfrak{X}^T w_1 = e_0, \quad \mathfrak{X}^T w_2 = e_n,$$

provided that the matrices P, Q of (2.15) (replacing l, l' by n and n by $2n$) are regular. In virtue of (2.25a), we have

$$z_1 = J_{2n}\, x_2, \quad z_2 = J_{2n}\, x_1,$$

$$w_1 = J_{2n}\, y_4 = \begin{bmatrix} 0 & I_n \\ I_n & 0 \end{bmatrix} x_2, \quad w_2 = J_{2n}\, y_3 = \begin{bmatrix} 0 & I_n \\ I_n & 0 \end{bmatrix} x_1.$$

We obtain that $\mathfrak{X}$ can be constructed, assuming P and Q are regular, by help of the two solutions x_1, x_2 of (2.24a).

We note finally that x_1 and x_2 can be expressed as

$$x_1 = \frac{1}{2}((u_1, J_n u_1) + (v_1, -J_n v_1)), \quad x_2 = \frac{1}{2}((u_2, J_n u_2) + (-v_2, J_n v_2)),$$

where u_k, v_k are the solutions of

$$A\, u_1 = e_0, \quad A\, u_2 = e_{n-1}, \quad A_-\, v_1 = e_0, \quad A_-\, v_2 = e_{n-1}.$$

A disadvantage of the inversion procedure sketched above consists in the additional assumption about the regularity of A_-. We shall now present a second approach, which can be applied to arbitrary regular T+H-matrices. This approach bases upon the fact that T+H-matrices have UV-rank ≤ 4 if one chooses

$$U = V = U_n + U_n^T.$$

Indeed, we have for $T = T_n(a)$, $H = H_n(b)$, $A = T + H$

$$A(U_n + U_n^T) - (U_n + U_n^T)A = \sum_{i=1}^{4} g_i\, f_i^T, \tag{2.26}$$

where

$$g_1 = e_0, \; g_2 = e_{n-1}, \; g_3 = -(a_1+b_{-1} \ldots a_n+b_{n-2}),$$
$$g_4 = -(a_{-n}+b_n \ldots a_{-1}+b_{2n-1}),$$
$$f_1 = (a_{-1}+b_{-1} \ldots a_{-n}+b_{n-2}), \; f_2 = (a_n+b_n \ldots a_1+b_{2n-1}),$$
$$f_3 = e_0, \quad f_4 = e_{n-1}.$$

Let x_i, y_i denote the solutions of

$$A\, x_i = g_i \qquad (i = 1,\ldots,4) \tag{2.27a}$$
$$A^T\, y_i = f_i \tag{2.27b}$$

and G, F, X, Y the matrices with columns g_i, f_i, x_i, y_i, respectively.

PROPOSITION 2.4. Suppose the equations (2.27a) are solvable, then A is regular and the columns c_k of the inverse can be computed successively by the recursion

$$c_{-1} := 0, \qquad c_0 = x_1,$$
$$c_{k+1} = (U_n + U_n^T - XF^T)c_k - c_{k-1} \tag{2.28a}$$

or

$$c_{k+1} = (U_n + U_n^T)c_k - c_{k-1} - XY^T e_k. \tag{2.28b}$$

PROOF. The completeness conditions (C1) and (C2) are satisfied. Consequently, the regularity of A follows by Theorem 1.1. According to (2.26), we have

$$(U_n + U_n^T - XF^T)\, A^{-1} = A^{-1}(U_n + U_n^T).$$

Hence,

$$A^{-1}\, U_n\, e_k = (U_n + U_n^T - XF^T)\, A^{-1} e_k - A^{-1}\, U_n^T\, e_k,$$

which implies (2.28a) and (2.28b). ≡

Let us note that the second inversion variant can be applied replacing the equations (2.27a) for $i = 3,4$ by

$$A z_1 = (k+1)_0^{n-1} \quad \text{and} \quad A z_2 = ((-1)^k(k+1))_0^{n-1},$$

because the vectors $(k+1)_0^{n-1}$ and $((-1)^k(k+1))_0^{n-1}$ are eigenvectors of $U_n + U_n^T$ (analogously for (2.27b), $i = 1,2$). The third inversion variant consists in replacing of (2.27a) for $i = 3,4$ by

$$A u_1 = e_1 \quad \text{and} \quad A u_2 = e_{n-2}.$$

2.3. Matrices close to Vandermonde. A further class of matrices, for which the inversion procedure via UV-reduction can be applied is the class of matrices close to Vandermonde. In this case we take as reduction operators $U = \mathrm{diag}(a_0 \dots a_{n-1})$ and $V = U_m^T$. An $m\times n$ matrix for which the UV-rank (with U,V chosen in such way) is small comparing with m and n will be called close to Vandermonde.

Let us give some examples. First of all, of course, we have to show that Vandermonde matrices are close to Vandermonde. As above, for $a = (a_0 \dots a_{n-1}) \in \mathbb{C}^n$ $V_m(a)$ denotes the $m\times n$ Vandermonde matrix

$$V_m(a) = \begin{bmatrix} 1 & 1 & \dots & 1 \\ a_0 & a_1 & \dots & a_{n-1} \\ \vdots & & & \\ a_0^{m-1} & a_1^{m-1} & \dots & a_{n-1}^{m-1} \end{bmatrix}. \tag{2.29}$$

Further, let diag a denote the diagonal matrix

$$\mathrm{diag}\, a := \mathrm{diag}(a_0 \dots a_{n-1}).$$

Then we have for $A = V_m(a)$

$$A\, \mathrm{diag}\, a - U_m^T A = e_{m-1}\, f^T, \tag{2.30}$$

where $f = (a_0^m \dots a_{n-1}^m)$. Obviously, the reduction (2.30) is complete if the numbers a_i are pairwise different, which means A is regular.

Matrices of the form

$$A = \begin{bmatrix} V_l(a) & \mathrm{diag}\, p \\ V_{m-l}(a) & \mathrm{diag}\, q \end{bmatrix}, \tag{2.31}$$

where $a, p, q \in \mathbb{C}^n$ will be called paired Vandermonde matrices. Such matrices occur when solving singular integral equations by the collocation method. We have here putting as above $U = \mathrm{diag}\, a$, $V = U_m^T$

$$AU - VA = e_{l-1}\, f_1^T + e_{m-1}\, f_2^T, \tag{2.32}$$

where

$$f_1 = (p_i\, a_i^{l} - q_i)_o^{n-1}, \quad f_2 = (q_i\, a_i^{m-1})_o^{n-1} .$$

That means A has UV-rank ≤ 2. In generalization of paired Vandermonde matrices one may consider m×n <u>matrices of V-mosaic type</u>

$$A = [V_{ik}]_{1\ 1}^{s\ t} ,$$

where $V_{ik} = V_{m_i}(a_k)$ diag p_{ik}. Choosing $U = \text{diag}(a_1 \dots a_t)$ and $V = U_m^T$ these matrices have UV-rank $\leq s$. Obviously, it is no restriction of generality to assume $t = 1$.

A further example of matrices close to Vandermonde are matrices of the form $A = [a_{ik}]_o^{m-1}{}_o^{n-1}$, where

$$a_{ik} = \sum_{s=o}^{\infty} c_{s-i}\, a_k^s \tag{2.33}$$

with $a = (a_o \dots a_{n-1}) \in \mathbb{C}^n$ and c_i $(i = -m, 1-m, \dots)$ complex numbers such that $\sum_i |c_i| < \infty$. Matrices of this form occur solving Wiener-Hopf equations by the collocation method. Choosing again $U = \text{diag } a$, $V = U_m^T$ we have

$$AU - VA = g\, f_1^T + e_{m-1}\, f_2^T, \tag{2.34}$$

where

$$f_1 = (1,1 \dots 1), \quad g = -(c_{-1} \dots c_{-m}), \quad f_2 = \left(\sum_{s=o}^{\infty} c_{s-m}\, a_k^s\right)_o^{n-1} .$$

Therefore, the UV-rank of A is ≤ 2.

Now we are going to construct, analogously to matrices close to Toeplitz, inversion formulae. Suppose A is an m×n matrix, such that

$$A \text{ diag } a - U_m^T A = \sum_{i=1}^{r} g_i\, f_i^T. \tag{2.35}$$

We shall assume that <u>the components of the vector a are pairwise distinct.</u> To guarantee that the completeness conditions (C1) and (C2) are satisfied one may take, as a normalization operator,

$$N := e_{m-1}\, 1_n^T,$$

where $1_n := (1,1 \dots 1) \in \mathbb{C}^n$. We construct inverses and one-side inverses of A using the solutions of the fundamental equations

$$A\, x_i = g_i, \quad A\, x_o = e_{m-1}, \tag{2.36a}$$

$$A^T y_i = f_i \quad (i = 1, \dots, r). \tag{2.36b}$$

We denote by X, Y, G, F the matrices with columns x_i, y_i, g_i, f_i $(i = 1, \dots, r)$, respectively.

<u>THEOREM 2.4.</u> Suppose A is an m×n matrix with (2.35), and the equations (2.36a) are solvable. Then rank $A = m$, and the matrix with

the columns c_k $(k = 0,\dots,m-1)$, where

$$c_{m-1} = x_o, \quad c_{k-1} = (\operatorname{diag} a - XF^T)c_k, \tag{2.37}$$

is a right inverse of A.

REMARK 2.4. In case A is square the recursion (2.37) can be replaced by

$$c_{k-1} = (\operatorname{diag} a)c_k - XY^T e_k.$$

THEOREM 2.5. Let A be a regular $n\times n$ matrix. Then the inverse of A can be represented in the following form

$$A^{-1} = (\operatorname{diag} x_o)\, V_n(a)^T J_n - \sum_{i=1}^{r} (\operatorname{diag} x_i)\, V_n(a)^T J_n T_n(U_n J_n y_i), \tag{2.38}$$

where x_i, y_i are the solutions of (2.36).

PROOF. In virtue of I, Lemma 1.1, (2.35) implies

$$A^{-1} e_{n-1-k} = (\operatorname{diag} a)^k x_o - \sum_{s=o}^{k-1} (\operatorname{diag} a)^s\, XY^T (U_m^T)^{k-1-s}\, e_{n-1}$$

which is equivalent to

$$A^{-1} J_n = (\operatorname{diag} x_o) V_n(a)^T - \sum_{i=1}^{r} (\operatorname{diag} x_i) V_n(a)^T \begin{bmatrix} y_n^{(i)} & \dots & y_1^{(i)} \\ & \ddots & \vdots \\ & & y_n^{(i)} \end{bmatrix},$$

where $y_i = (y_k^{(i)})_o^{n-1}$, $y_n^{(i)} := 0$. This is another expression of (2.38) and completes the proof. ≡

COROLLARY 2.1. The inverse of a regular Vandermonde matrix $V_n(a)$ can be expressed in the form

$$V_n(a)^{-1} = -(\operatorname{diag} x)\, V_n(a)^T \begin{bmatrix} \eta_1 & \dots & \eta_n \\ \vdots & ⋰ & \\ \eta_n & & \end{bmatrix}, \tag{2.39}$$

where $\eta_n := -1$ and $x, y := (\eta_i)_o^{n-1}$ are the solutions of

$$V_n(a)x = e_o, \quad V_n(a)^T y = (a_i^n)_o^{n-1}.$$

In the case of matrices close to Toeplitz we could determine left inverses going over to the transposed matrix and computing a right inverse of A^T by Theorem 2.1. This procedure is not applicable to matrices close to Vandermonde because of the "non-symmetric" choice of the reduction operators. Therefore, we note a separate theorem for constructing left inverses. Additionally to (2.36b) we consider the equation

$$A^T y_o = 1_n. \tag{2.40}$$

THEOREM 2.6. Suppose the equations (2.36b), (2.40) are solvable, then rank $A = n$. Furthermore, if P denotes the matrix with rows p_k^T $(k = 0,\dots,n-1)$, where

$$p_0 = y_0, \quad p_{k+1} = (U_m + YG^T)p_k,$$

then $V_n(a)^{-1}P$ is a left inverse of A.

PROOF. The relation rank $A = n$ is obtained applying Prop. 1.3. Furthermore, in virtue of (2.35), we have

$$A \text{ diag } a = (U_m^T + GY^T)A,$$

which implies

$$1_n^T(\text{diag } a)^k = y_0^T\, A(\text{diag } a)^k = y_0^T(U_m^T + GY^T)^k A.$$

The latter relation can be expressed as

$$V_n(a) = PA,$$

which completes the proof. ≡

As for matrices close to Toeplitz the second inversion variant can be applied. Let us explain this for the simplest case, the case of a square Vandermonde matrix $A = V_n(a)$. Instead of the equation $A^T y = f$, where $f = (a_i^n)_0^{n-1}$ we consider one of the equations

$$A^T w_l = e_l,$$

since e_l is an eigenvector of $U^T = \text{diag } a$. Clearly, w_l^T is the l th row of A^{-1}. The solution w_l is connected with y via the formula

$$y = -\frac{1}{e_{n-1}^T w_l}(U_n - a_l I_n)w_l,$$

because, by Cramer's rule, the last component of w_l does not vanish. We have obtained that A^{-1} can be constructed by means of the last column and the l th row.

Finally in this subsection we discuss the problem how to establish A knowing $D_A = A \text{ diag } a - U_m^T A$, where $a = (a_i)_0^{n-1}$. We note that there is an essential difference between matrices close to Toeplitz and matrices close to Vandermonde basing on the fact that on one side the reduction operators in the close to Toeplitz case have 0 as a joint point of their spectrum and on the other side in the close to Vandermonde case the spectra of U and V do not intersect, provided that $a_i \neq 0$.

PROPOSITION 2.5. Suppose A is an $m\times n$ matrix such that (2.35) holds, where $a_i \neq 0$. Then

$$c^T := e_{m-1}^T A = e_{m-1}^T\, GF^T(\text{diag } a)^{-1} \tag{2.41}$$

and

$$A = J_m V_m(a^{-1}) \text{diag } c + J_m \sum_{i=1}^{r} T_m(g_i') \, V_m(a^{-1}) \text{diag } f_i (\text{diag } a)^{-1}, \quad (2.42)$$

where $g_i' = g_i - (e_o^T g_i) e_o$ and $V_m(a^{-1}) = V_m(a_1^{-1} \ldots a_n^{-1})$.

PROOF. We have, by assumption,

$$e_k^T A \text{ diag } a - e_{k+1}^T A = e_k^T GF^T.$$

Denoting the rows of the matrix $J_m A$ by c_k^T this implies $c_o = c$ and

$$c_{k+1}^T = c_k^T (\text{diag } a)^{-1} + e_{k+1}^T GF^T (\text{diag } a)^{-1}.$$

We conclude

$$c_k^T = c^T (\text{diag } a)^{-k} + \sum_{s=1}^{k} e_s^T GF^T (\text{diag } a)^{-k-1+s}$$

for $k = 0,\ldots,m-1$, which yields (2.42). ≡

<u>2.4. Matrices with a small rank diagonal reduction.</u> In this subsection we deal with UV-Toeplitz-like matrices if U and V are diagonal matrices, i.e. $m \times n$ matrices A, for which

$$A \text{ diag } b - (\text{diag } c) A = \sum_{j=1}^{r} g_j f_j^T, \quad (2.43)$$

where r is small comparing with m and n. We always shall assume that the components of each of the vectors $c = (c_i)_o^{n-1}$ and $b = (b_i)_o^{m-1}$ are pairwise distinct. In this case for all i there is at most one k such that $c_i = b_k$ and, vice versa, for all k there is at most one i with this property. Therefore, the general case can be led after changing the order of columns and rows of the matrices diag a, diag b and A to the case, where <u>$c_i = b_k$ at most for $i = k$.</u>

Let us give now some examples. Suppose $A = [a_{ik}]$, where

$$a_{ik} = \frac{1}{b_k - c_i} \quad (c_i \neq b_k) \; ; \quad c = (c_i)_o^{m-1}, \quad b = (b_i)_o^{n-1}. \quad (2.44)$$

Then putting $U = \text{diag } b$ and $V = \text{diag } c$ we have

$$AU - VA = \mathbb{1}_m \mathbb{1}_n^T, \quad (2.45)$$

where $\mathbb{1}_k := (1,1\ldots1) \in \mathbb{C}^k$. This UV-reduction is, by assumption, complete. A further class of matrices admitting a representation (2.43) are matrices of the form

$$A = V_1(c)^T J_1 V_1(b), \quad (2.46)$$

where $c \in \mathbb{C}^m$, $b \in \mathbb{C}^n$. Choosing U, V as above we obtain after an elementary computation

$$AU - VA = \mathbb{1}_m f_b^T - f_c \mathbb{1}_n^T, \tag{2.47}$$

where $f_c = (c_i^1)_{i=o}^{m-1}$, $f_b = (b_i^1)_{i=o}^{n-1}$, which is a complete reduction with UV-rank 2. In generalization of matrices (2.46) one may consider matrices of the form

$$A = V_1(c)^T H V_1(b), \tag{2.48}$$

where H is an lxl Hankel matrix. We recall that the problem of computing the Moore-Penrose inverse of an H-matrix can be led to the problem of inverting a matrix (2.48) (see I, Subsection 7.4, Prop. 7.1).
It is easily seen that matrices of the form (2.48) have a UV-rank ≤ 4, if U and V are defined as above.
We note that a matrix A can easily be established knowing $AU - VA$ and $A \mathbb{1}_n$.

<u>PROPOSITION 2.6.</u> Suppose $A = [a_{ik}]$ is an mxn matrix and $A \mathbb{1}_n = (s_i)_o^{m-1}$. Then (2.43) holds with $g_j = (g_i^{(j)})_o^{m-1}$, $f_j = (f_i^{(j)})_o^{n-1}$ $(j=1,\dots,r)$ if and only if

$$a_{ik} = -\frac{1}{b_i - c_k} \sum_{j=1}^{r} g_i^{(j)} f_k^{(j)} \quad (i \neq k), \quad a_{ii} = s_i - \sum_{k \neq i} a_{ik}.$$

The proof is obvious. ≡

The UV-reduction (2.43) is, clearly, not necessarily complete. A suitable choice of a normalization operator is in this case

$$N := \mathbb{1}_m \mathbb{1}_n^T .$$

Analogously to matrices close to Toeplitz and Vandermonde we construct inverses and one-side inverses using the solutions of the fundamental equations

$$A x_i = g_i , \quad A x_o = \mathbb{1}_m , \tag{2.49a}$$

$$A^T y_i = f_i \quad (i = 1,\dots,r). \tag{2.49b}$$

As above we denote by X, F the matrix with columns x_i, f_i, respectively.

<u>PROPOSITION 2.7.</u> Suppose the equations (2.49a) are solvable, then rank $A = m$. Furthermore, if P denotes the matrix with columns p_k $(k = 0,\dots,m-1)$, where

$$p_o = x_o, \quad p_{k+1} = (\operatorname{diag} b - XF^T)p_k,$$

then $P(V_m(c)^T)^{-1}$ is a right inverse of A.

<u>PROOF.</u> Taking into account that (C1) is fulfilled the first assertion immediately follows from Prop. 1.3'. Furthermore, one concludes from (2.43) that

$$Ap_k = A(\operatorname{diag} b - XF)^k x_o = (\operatorname{diag} c)^k \mathbb{1}_m,$$

which implies $AP = V_m(c)^T$. ≡

THEOREM 2.7. Let A be a regular $n \times n$ matrix. Then the inverse $A^{-1} = [c_{ik}]_0^{n-1}$ can be represented as follows

$$c_{ik} = \frac{1}{b_i - c_k} \sum_{j=1}^{r} x_i^{(j)} y_k^{(j)} \quad (i \neq k), \quad c_{ii} = x_i^{(0)} - \sum_{k \neq i} c_{ik},$$

where $x_j = (x_i^{(j)})_0^{n-1}$, $y_j = (y_i^{(j)})_0^{n-1}$ are the solutions of (2.49).

Let us still note that the second inversion variant can be applied. We explain this for matrices of the form (2.44). Suppose z is the solution of

$$Az = e_0$$

and $1_n^T z \neq 0$, then the solution of the fundamental equation

$$A\, x_0 = 1_n$$

is given by

$$x_0 = \frac{1}{1_n^T z} \left((\operatorname{diag} c)z - b_0 z \right),$$

where b_0 is the first component of b.

3. INVERSION ALGORITHMS FOR TOEPLITZ-LIKE MATRICES

To obtain inversion algorithms for general Toeplitz-like matrices we proceed analogously to Toeplitz and Hankel matrices in Part I. That means, after presenting inversion formulae using the fundamental solutions we are going now to construct recursion formulae. This leads to inversion algorithms for strongly regular Toeplitz-like matrices, which require $O(n^2)$ operations and linear storage. Let us note that the construction of the recursion formulae for all special types of Toeplitz-like matrices (close to Toeplitz, close to Hankel, close to Vandermonde, with a small rank diagonal reduction) needs specified investigation.

3.1. General recursion formulae for the fundamental equations.

In this subsection we consider general $n \times n$ UV-Toeplitz-like matrices A for reduction operators $U = [u_{ik}]_0^{n-1}$ and $V = [v_{ik}]_0^{n-1}$ satisfying the following condition

$$u_{ik} = 0 \quad (i-k > 1), \quad v_{ik} = 0 \quad (k-i > 1). \tag{3.1}$$

The principal sections of U and V are denoted by $(U)_m$ and $(V)_m$, $(U)_m = [u_{ik}]_0^{m-1}$, $(V)_m = [v_{ik}]_0^{m-1}$. (We use parentheses to distinguish between the sections of U and the shift in $\mathbb{C}^m$.) First we study how the UV-reduction of a Toeplitz-like matrix $A = [a_{ik}]_0^{n-1}$ is connected with the $(U)_m(V)_m$-reduction of its principal sections $A_m = [a_{ik}]_0^{m-1}$.

LEMMA 3.1. Suppose

$$AU - VA = \sum_{i=1}^{r} g_i f_i^T, \tag{3.2}$$

where U and V satisfy (3.1). Then

$$A_m(U)_m - (V)_m A_m = \sum_{i=0}^{r+1} g_i^m (f_i^m)^T, \tag{3.3}$$

with

$$\begin{aligned} g_i^m &= I_{mn}^o g_i, \quad f_i^m = I_{mn}^o f_i \quad (i = 1,\dots,r), \\ g_0^m &= (a_{im})_0^{m-1}, \quad f_0^m = \varrho_m e_{m-1}, \quad \varrho_m := -u_{m,m-1}, \\ f_{r+1}^m &= (a_{mk})_0^{m-1}, \quad g_{r+1}^m = \sigma_m e_{m-1}, \quad \sigma_m := v_{m-1,m}. \end{aligned} \tag{3.4}$$

PROOF. The equality (3.3) is obtained multiplying the equality (3.2) by I_{mn}^o from the left and I_{nm}^o from the right. ≡

From (3.3) it becomes clear that we have to consider the following "fundamental equations"

$$A_m x_i^m = g_i^m, \tag{3.5a}$$

$$(i = 0,\dots,r+1)$$

$$A_m^T y_i^m = f_i^m. \tag{3.5b}$$

Our aim is to construct formulae for the recursion $x_i^m \to x_i^{m+1}$ and $y_i^m \to y_i^{m+1}$. We shall see that the construction is quite simple and do not require any further information if we consider the equations (3.5a) for $i = 1,\dots,r+1$ and (3.5b) for $i = 0,\dots,r$. Whereas for the remaining two equations it is necessary to specify our investigation, which will be done in the forthcoming subsections.

The construction of recursion formulae mainly bases upon the following property.

LEMMA 3.2. Suppose $A = [a_{ik}]_0^m$ is regular and $A_m = [a_{ik}]_0^{m-1}$ is such that the equations

$$A_m x_0^m = g_0^m \quad \text{and} \quad A_m x^m = g \qquad (g \in \mathbb{C}^m)$$

are solvable. Then the solution of the equation

$$Ax = \begin{bmatrix} g \\ \gamma \end{bmatrix} \qquad (\gamma \in \mathbb{C})$$

is given by

$$x = \begin{bmatrix} x^m \\ 0 \end{bmatrix} + \frac{\gamma - \beta_m}{\alpha_m} \begin{bmatrix} x_o^m \\ -1 \end{bmatrix}, \tag{3.6}$$

where

$$\alpha_m = (f_{r+1}^m)^T x_o^m - a_{mm} \neq 0, \quad \beta_m = (f_{r+1}^m)^T x^m.$$

PROOF. We have $\alpha_m \neq 0$, since otherwise $\begin{bmatrix} x_o^m \\ -1 \end{bmatrix}$ is a non-trivial element of ker A, which contradicts our assumption. The equality (3.6) is verified by a direct calculation. ≡

From Lemma 3.2 immediately follow recursion formulae for the fundamental equations.

PROPOSITION 3.1. Suppose A_m and A_{m+1} are regular principal sections of a matrix $A = [a_{ik}]$ with a UV-reduction (3.2). Then the solutions of the equations (3.5a) fulfil the recursion

$$x_i^{m+1} = \begin{bmatrix} x_i^m \\ 0 \end{bmatrix} + \frac{\gamma_{im}}{\alpha_m} \begin{bmatrix} x_o^m \\ -1 \end{bmatrix} \quad (i = 1,\ldots,r), \tag{3.7a}$$

$$x_{r+1}^{m+1} = \frac{\sigma_{m+1}}{\alpha_m} \begin{bmatrix} x_o^m \\ -1 \end{bmatrix}, \tag{3.7b}$$

where

$$\alpha_m = (f_{r+1}^m)^T x_o^m - a_{mm} \neq 0 \quad \text{and} \quad \gamma_{im} = e_m^T g_i^m - (f_{r+1}^m)^T x_i^m.$$

Obviously, the solutions of (3.5b) admit a similar recursion replacing x_o^m by y_{r+1}^m and f_{r+1}^m by g_o^m.

3.2. Inversion of matrices close to Toeplitz. Let us specify our investigation to the case $U = V = U_n$, where U_n is, as above, the forward shift in $\mathbb{C}^n$. Our aim is to obtain a recursion formula for the solution x_o^m of (3.5a) (i = 0). For this we make use of the connection (3.7b) between x_o^m and the solution of

$$A_{m+1} z^{m+1} = e_m.$$

As easily seen, this connection is given by

$$z^{m+1} = \frac{1}{\alpha_m} \begin{bmatrix} x_o^m \\ -1 \end{bmatrix}. \tag{3.8}$$

To construct x_o^{m+1} we apply the second inversion variant. We have

$$A_{m+1} U_{m+1} z^{m+1} = \sum_{i=o}^{r} g_i^{m+1} (f_i^{m+1})^T z^{m+1},$$

since $g_{r+1}^{m+1} = 0$ and $e_m \in \ker U_{m+1}$. Hence,

$$U_{m+1} z^{m+1} = \sum_{i=o}^{r} x_i^{m+1} (f_i^{m+1})^T z^{m+1}.$$

The latter relation together with (3.8) leads to the following.

PROPOSITION 3.2. Suppose A_m and A_{m+1} are regular sections of A with the UV-reduction (3.2), where $U = V = U_n$. Then

$$x_o^{m+1} = \begin{bmatrix} 0 \\ x_o^m \end{bmatrix} - \sum_{i=1}^{r} x_i^{m+1}(f_i^{m+1})^T \begin{bmatrix} x_o^m \\ -1 \end{bmatrix}. \tag{3.9}$$

The recursion formulae (3.7a) and (3.9) together with inversion formulae (2.9) lead to an algorithm for inverting strongly regular matrices closed to Toeplitz. The computational complexity of this algorithm is about $(4r + \frac{1}{2})n^2$ multiplications.

If we want to apply the inversion formula (2.10) we have to evaluate the vectors y_i $(i = 0,\dots,r)$, too. Unfortunately the evaluation of these solutions is not completely analogous to that one of the vectors x_i. To establish an algorithm for the determination of y_i we study both the U_nU_n-reduction and the $U_n^TU_n^T$-reduction of the matrix A and write these in the following form

$$AU_m - U_nA = -g_{r+1}e_{n-1}^T + e_o f^T + \sum_{i=1}^{r} \begin{bmatrix} 0 \\ g_i \end{bmatrix}[f_i^T\ 0], \tag{3.10a}$$

$$AU_n^T - U_n^TA = -g\,e_o^T + e_{n-1}f_{r+1}^T - \sum_{i=1}^{r} \begin{bmatrix} g_i \\ 0 \end{bmatrix}[0\ f_i^T], \tag{3.10b}$$

where $g_{r+1} := (a_{i-1,n-1})_o^{n-1}$, $g := (a_{io})_1^n$, $f := (a_{oi})_1^n$, $f_{r+1} := (a_{n-1,i-1})_o^{n-1}$ with $a_{ik} = 0$ if $i, k \notin \{0,1,\dots,n-1\}$.

Now let us introduce fundamental solutions corresponding to the reduction (3.10a)

$$A^T y_o = f, \quad A^T y_i = \begin{bmatrix} f_i \\ 0 \end{bmatrix} \quad (i = 1,\dots,r), \quad A^Tw = e_{n-1}, \tag{3.11}$$

where the vector w can be determined as z^n due to (3.8).

The evaluation of the solutions y_i $(i = 0,\dots,r)$ is divided into two steps:

1st recursive computation of the solutions $\tilde{y}_i^m$, $\tilde{y}^m$ $(m = 1,\dots,n)$ of

$$A_m^T \tilde{y}_i^m = I_{mn}^o \begin{bmatrix} 0 \\ f_i \end{bmatrix} \quad (i = 1,\dots,r),$$

$$A_m^T \tilde{y}_o^m = (a_{mi})_o^{m-1}, \quad A_m^T \tilde{y}^m = e_o,$$

2nd evaluation of the solutions y_i $(i = 0,\dots,r)$ by means of the solutions

$$\tilde{y}_k := \tilde{y}_k^n \quad (k = 0,\dots,r).$$

Replacing A by A^T and using the transposed equality of (3.10b) the first step is realized analogously to the computation of x_i^m.

Now we discuss the realization of the second step. Obviously, we have, according to the transposed equality of (3.10a),

$$A^T U_n^T \tilde{y}_i = [{f_i \atop 0}] + e_{n-1}\, g_{r+1}^T \tilde{y}_i - f e_o^T \tilde{y}_i - \sum_{k=1}^{r} [{f_k \atop 0}][0\ g_k^T]\, \tilde{y}_i$$

for $i = 0,\dots,r$. This implies

$$U_n^T \tilde{y}_i = y_i + w\, g_{r+1}^T \tilde{y}_i - y_o\, e_o^T \tilde{y}_i - \sum_{k=1}^{r} y_k\, [0\ g_k^T]\, \tilde{y}_i .$$

Defining numbers

$$\alpha_{ik} = \delta_{ik} - [0\ g_k^T]\, \tilde{y}_i \quad (i = 0,\dots,r;\ k = 1,\dots,r),$$

$$\alpha_{io} = -\, e_o^T \tilde{y}_i\, \delta_{io}$$

the latter relation is equivalent to

$$\sum_{k=o}^{r} \alpha_{ik}\, y_k = (U_n^T - w\, g_o^T)\, \tilde{y}_i . \tag{3.12}$$

Solving this $(r+1)\times(r+1)$ system y_k are obtained.
Finally, let us remark that the reason behind this somewhat more complicate recursion is that for $U = V = U_n$ in the reduction (3.3) δ_m vanishes. This is obviously different to the case of $U_n U_n^T$-Toeplitz-like matrices, which are matrices close to Hankel. For a matrix A of this class it is possible to realize the recursion of the fundamental solutions for A^T similarly to that one for A (due to Prop. 3.1, 3.2). Here the role of z^n will play the solution x^n of $A\, x^n = e_o$ (because $e_o \in \ker U_n^T$), which is also recursively determined by means of Prop. 3.1.

<u>3.3. Inversion of matrices close to Vandermonde.</u> The second special case under consideration is $U = \operatorname{diag} a$, $a = (a_i)_o^{n-1}$, $V = U_n^T$ (c.f. Subsection 2.3). For convenience we write the reduction (3.3) of the principal sections A_m as follows

$$A_m D_m - U_m^T A_m = G_m F_m^T + e_{m-1}(f_{r+1}^m)^T = \tilde{G}_m \tilde{F}_m^T \tag{3.13}$$

where $D_m := \operatorname{diag} a^m$, $a^m := I_{mn}^o a$ and G_m, F_m denote the matrices with the columns $I_{mn}^o g_i$, $I_{mn}^o f_i$ $(i = 1,\dots,r)$, respectively $\tilde{G}_m := [G_m\ e_{m-1}]$, $\tilde{F}_m := [F_m\ f_{r+1}^m]$. First we observe that in the strongly regular case the solutions y_i^m of (3.5b) can be successively computed by Prop. 3.1, 3.2, and the inverse of A is obtained applying Theorem 2.6. The inversion formula (2.38) requires the evaluation of the solutions x_i^m of (3.5a), too. As discussed in Subsection 3.1, for this it remains to show how to determine x_o^m $(m = 1,\dots,n)$. First we give a representation of the vector g_o^m. We have

$$(A_{m+1}D_{m+1} - U^T_{m+1}A_{m+1})e_m = (a_m I_{m+1} - U^T_{m+1})A_{m+1}\, e_m$$

$$= G_{m+1}F^T_{m+1}\, e_m + e^T_m\, f^{m+1}_{r+1}\, e_m.$$

Since $g^m_o = I^o_{m,m+1}A_{m+1}e_m$ and $I^o_{m,m+1}(a_m I_{m+1} - U^T_{m+1})^{-1}e_m =$ $= a_m^{-1}(a_m I_m - U^T_m)^{-1}e_m$ we obtain

$$g^m_o = (a_m I_m - U^T_m)^{-1}\,\tilde{G}_m\,\Phi_m\,,$$

where

$$\Phi_m := \begin{bmatrix} F^T_{m+1}\, e_m \\ a_m^{-1}\, e^T_m\, \tilde{G}_{m+1}\, \tilde{F}^T_{m+1}\, e_m \end{bmatrix}.$$

PROPOSITION 3.3. Suppose A_m is a regular section of A. Then the matrix

$$\theta(a_m) = I_{r+1} + \tilde{F}_m(a_m I_m - D_m)^{-1}\,\tilde{X}_m$$

is regular and

$$x^m_o = (a_m I_m - D_m)^{-1}\,\tilde{X}_m\,\theta(a_m)^{-1}\,\Phi_m. \tag{3.14}$$

PROOF. From (3.13) one concludes that the matrix $K := a_m I_m - D_m + \tilde{X}_m\tilde{F}^T_m$ is regular and

$$A_m K^{-1} = (a_m I_m - U^T_m)^{-1}\,A_m.$$

Hence,

$$A_m K^{-1}\,\tilde{X}_m\,\Phi_m = (a_m I_m - U^T_m)^{-1}\,\tilde{G}_m\,\Phi_m = g^m_o,$$

which is $x^m_o = K^{-1}\,\tilde{X}_m\,\Phi_m$. To compute K^{-1} we use the formula (c.f. Subsection 1.3)

$$K^{-1} = (a_m I_m - D_m)^{-1}(I_m - \tilde{X}_m\theta(a_m)^{-1}\,\tilde{F}^T_m(a_m I_m - D_m)^{-1})$$

and the fact that the regularity of K implies the regularity of $\theta(a_m)$. Taking into account the relation

$$I_{r+1} - \theta(a_m)^{-1} = \theta(a_m)^{-1}\,\tilde{F}_m(a_m I_m - D_m)^{-1}\,\tilde{X}_m \tag{3.15}$$

we obtain the equality (3.14). ≡

3.4. Inversion of matrices with a small rank diagonal reduction.

Suppose $A = [a_{jk}]^{n-1}_o$ is a regular nxn matrix such that (2.43) holds. Using the notations of Subsection 3.3 we obtain for the principal sections A_m

$$A_m \operatorname{diag} b^m - (\operatorname{diag} c^m)A_m = G_M F_m^T. \tag{3.16}$$

We have to consider the following fundamental equations

$$A_m x_i^m = g_i^m \quad (i = 1,\dots,r), \quad A_m x^m = \mathbf{1}_m, \tag{3.17a}$$

$$A_m x_o^m = (a_{jm})_o^{m-1} =: g_o^m \tag{3.17b}$$

and the corresponding equations for the transposed matrix. In the strongly regular case the solutions of (3.17a) can be determined due to Prop. 3.1. Now we are going to deduce a recursion formula for the vector x_o^m. From relation (3.16) we conclude that the matrix $K := b_m I_m - \operatorname{diag} b^m + X_m F_m^T$ is regular and

$$A_m K^{-1} = (b_m I_m - \operatorname{diag} c^m)^{-1} A_m.$$

Hence,

$$\begin{aligned} A_m K^{-1} X_m F_m^T e_{m-1} &= (b_m I_m - \operatorname{diag} c^m)^{-1} G_m F_m^T e_{m-1} \\ &= \Big(\sum_{i=1}^{r} \frac{1}{b_m - c_j}(g_j^i)(f_m^i)\Big)_{j=o}^{m-1} = g_o^m , \end{aligned}$$

where $g_j^i = e_j^T g_i$, $f_m^i = e_m^T f_i$. This implies

$$x_o^m = K^{-1} X_m F_m^T e_{m-1}.$$

With analogous consideration as in the proof of Prop. 3.3 we obtain the following rule.

PROPOSITION 3.4. Suppose A_m is a regular section of A. Then the matrix

$$\theta(b_m) = I_r + F_m (b_m I_m - \operatorname{diag} b^m)^{-1} X_m$$

is regular and

$$x_o^m = (b_m I_m - \operatorname{diag} b^m)^{-1} X_m \theta(b_m)^{-1} F_m^T e_{m-1}. \tag{3.18}$$

Finally, let us consider, as an example, matrices $A = [a_{jk}]$, for which $a_{jk} = (b_k - c_j)^{-1}$ if $j \neq k$. We introduce the vector

$$\psi := ((b_m - b_k)^{-1})_o^{m-1} .$$

Then we have

$$\theta(b_m) = 1 + \mathbf{1}^T (b_m I_m - \operatorname{diag} b^m)^{-1} x^m = 1 + \psi^T x_m.$$

Therefore

$$x_o^m = \frac{1}{1 + \psi^T x^m} (b_m I_m - \operatorname{diag} b^m)^{-1} x^m.$$

3.5. A modification of the recursion. The recursion algorithms presented above can be only applied if the matrix is strongly regular. In some important special cases - such as paired Toeplitz and Vandermonde matrices - this cannot be guaranteed, but, on the other hand, it can be concluded (from the theory of projection methods) that the submatrices $A_{(m)} = [a_{ik}]_p^{n-1-p}$, $m = n - 2p$, of $A = [a_{ik}]_o^{n-1}$ are regular.
We have the following modification of Lemma 3.2.

LEMMA 3.3. Let $A_{(m)}$, $A_{(m+2)}$ be regular submatrices of A, $g^m \in \mathbb{C}^m$, $g, \tilde{g} \in \mathbb{C}$ and x^m, x^{m+2} the solutions of

$$A_{(m)}x^m = g^m , \quad A_{(m+2)}x^{m+2} = \begin{bmatrix} \tilde{g} \\ g^m \\ g \end{bmatrix} .$$

Then

$$x^{m+2} = \begin{bmatrix} 0 \\ x^m \\ 0 \end{bmatrix} + (\tilde{g} - \tilde{h}^T x^m)\, x_o^{m+2} + (g - h^T x^m)\, x_{m+1}^{m+2} ,$$

where $h = (a_{p-1,k+p})_o^{m-1}$, $\tilde{h} = (a_{n-p,k+p})_o^{m-1}$, $x_k^{m+2} = A_{(m+2)}^{-1}e_k$ $(k = 0,m+1)$.

Recursion formulae for the solutions x_k^{m+2} $(k = 0,m+1)$ are obtained using both the U_mU_m- and the $U_m^TU_m^T$-reduction in the paired Toeplitz case, and the $\tilde{D}_mU_m^T$- and the $\tilde{D}_m^{-1}U_m$-reduction (with $D_m:=\mathrm{diag}(a_p\ldots a_{n-1-p})$) in the paired Vandermonde case.

4. INVERSION OF INTEGRAL OPERATORS WITH DISPLACEMENT KERNEL AND THEIR GENERALIZATIONS

4.1. Preliminaries. In this section we consider some classes of operators acting within the Banach space $L_p := L_p[0,\tau]$ $(1 \le p < \infty)$ of all measurable functions $x(t)$ on a finite interval $[0,\tau]$, for which $|x(t)|^p$ is summable with the usual norm

$$\|x(t)\|_p := (\int_0^\tau |x(t)|^p\, dt)^{1/p}.$$

In all what follows we shall identify the space L_p with the subspace of $L_p(-\infty,\infty)$ consisting of all functions vanishing outside the interval $[0,\tau]$. That means, for $\varphi \in L_p$ we put $\varphi(t) = 0$ if $t < 0$ or $t > \tau$. Provided that $f \in L_1$ and $\varphi \in L_p$, we denote the convolution of f and φ by

$$(f*\varphi)(t) := \int_0^t f(t-s)\, \varphi(s)ds \quad (0 \le t \le \tau).$$

It is well-known that the operator $F_f\varphi \to f*\varphi$ is compact in L_p and $\|F_f\|_p \leq \|f\|_1$. In particular, $J\varphi := 1*\varphi$ [1] is the integration operator

$$(J\varphi)(t) = \int_0^t \varphi(s)ds.$$

Obviously, the operator $I - \mu J$ is invertible for all $\mu \in \mathbb{C}$ and

$$(I-\mu J)^{-1}\varphi = (\delta+\mu e^{\mu t})*\varphi = \varphi(t) + \int_0^t e^{\mu(t-s)}\varphi(s)ds, \tag{4.1}$$

where δ denotes the Dirac distribution. In particular, there is

$$(I - \mu J)^{-1}1 = e^{\mu t}. \tag{4.2}$$

We shall identify the dual space of L_p with L_q, $\frac{1}{p} + \frac{1}{q} = 1$, via the bilinear form

$$(f,g) := \int_0^\tau f(t)g(t)dt \quad (f \in L_p,\ g \in L_q).$$

The adjoint operator of J is given by

$$J'\varphi = \int_0^\tau \varphi(t)dt - J\varphi.$$

Clearly, $(\varphi, e^{\mu t})$ is the Laplace transformation of $\varphi(t) \in L_1$, which will be designated by $\hat{\varphi}(\mu)$. We make use of the following elementary well-known properties.

LEMMA 4.1.

1) $\hat{\varphi}(\lambda)\,\hat{\psi}(\lambda) = \widehat{\varphi * \psi}(\lambda)$

2) If $\hat{\varphi}(\lambda) = \lambda\hat{\psi}(\lambda)$, where $\varphi,\psi \in L_1$, then ψ is absolutely continuous and $\psi'(t) = -\varphi(t)$, $(\varphi,1) = 0$.

In the following we shall use the fact that any linear bounded operator in L_p can be represented as an integral operator. More precisely speaking, the following assertion holds true.

LEMMA 4.2. Let A be a linear bounded operator in L_p, then there is a function $a(t,s)$ such that $a_t(s) := a(t,s)$ belongs to L_q for almost all $t \in [0,\tau]$, $a(0,s) = 0$ and

$$(A\varphi)(t) = \frac{d}{dt}\int_0^\tau a(t,s)\,\varphi(s)ds \quad (0 \leq t \leq \tau). \tag{4.3}$$

PROOF. We have

$$A\varphi = \frac{d}{dt}\, J\varphi = \frac{d}{dt}\,(A\varphi,\theta_t(s)) = \frac{d}{dt}\,(\varphi,A'\theta_t(s)),$$

where $\theta_t(s) := \theta(t-s)$ and $\theta(\cdot)$ denotes the Heaviside function. Setting $a(t,s) := A'\theta_t(s)$ we obtain (4.3). ≡

4.2. The class of JJ-Toeplitz-like operators. According to the general definition of a Toeplitz-like operator, a JJ-Toeplitz-like operator

[1] In all what follows we do not distinguish between the number 1 and the function $\varphi(t) = 1$.

is an operator, for which $D_A := AJ - JA$ has finite rank. The class of JJ-Toeplitz-like operators in L_p, for which rank $D_A \leq r$ will be denoted by $\mathcal{T}^p_r$. In this subsection we shall give a general description of the class $\mathcal{T}^p_r$; in the following two subsections we shall consider some special cases.

For an operator $A \in \mathcal{T}^p_r$ there are functions $g_i \in L_p$ and $f_i \in L_q$ such that

$$D_A = (AJ - JA)\varphi = \sum_{i=1}^{r} (\varphi, f_i) g_i. \tag{4.4}$$

We ask whether the reduction is complete (c.f. Definition 1.2).

<u>PROPOSITION 4.1.</u> The reduction (4.4) is complete if and only if A is Fredholm.

<u>PROOF.</u> If (4.4) is complete then A is Fredholm, by Prop. 1.2. Now we assume that A is Fredholm. Further, let us assume for a moment that $\ker \mathrm{Obs}_k A \neq \{0\}$ for all $k = 0,1,2\ldots$. Then there exists a $u \in L_p$, $u \neq 0$ such that $J^k u \in \ker A$ for $k = 0,1,\ldots$. Since, by assumption, $\dim \ker A < \infty$, we have $\sum_{k=0}^{N} \alpha_k J^k u = 0$ for some α_k with $\alpha_N \neq 0$. This implies $u = 0$. The contradiction shows that $\ker \mathrm{Obs}_N A = \{0\}$. Analogously, one obtains $\ker(\mathrm{Con}_N A)' = \{0\}$. Because $\mathrm{Con}_N A$ is normally solvable, this implies $\mathrm{im}\, \mathrm{Con}_N A = L_p$.

According to the definition of Section 1, the following equations are fundamental

$$Ax_i = g_i, \quad (i = 1,\ldots,r) \tag{4.5a}$$

$$A' y_i = f_i. \tag{4.5b}$$

Applying Theorem 1.1 we obtain

<u>PROPOSITION 4.2.</u> Suppose $A \in \mathcal{T}^p_r$ is Fredholm, then the solvability of (4.5) implies the invertibility of A. Moreover, the inverse of A belongs to $\mathcal{T}^p_r$ again.

<u>REMARK 4.1.</u> If A has, additionally, the index zero, then the invertibility already follows from the solvability of (4.5a) (or (4.5b)).

Next we show how to construct A and A^{-1} knowing D_A, $D_{A^{-1}}$, respectively.

<u>THEOREM 4.1.</u> Suppose $A \in \mathcal{T}^p_r$, then (4.4) holds if and only if A is given by (4.3), where

$$a(t,s) = g_o(t-s) - \frac{\partial}{\partial s} \sum_{i=1}^{r} \int_{o}^{\min(t,s)} g_i(t-\sigma) f_i(s-\sigma)\, d\sigma \tag{4.6a}$$

and

$$g_o = A1. \tag{4.6b}$$

For the proof we need the following assertion.

<u>LEMMA 4.3.</u> The relation (4.4) is equivalent to

$$Ae^{\mu t} = (I-\mu J)^{-1}g_0 + \mu \sum_{i=1}^{r}(I-\mu J)^{-1}g_i(t)\hat{f}_i(\mu). \tag{4.7}$$

PROOF. First let us assume that (4.4) holds. Then we have for all $\mu \in \mathbb{C}$

$$A(I-\mu J)^{-1}\varphi = (I-\mu J)^{-1}A\varphi + \mu(I-\mu J)^{-1} \sum_{i=1}^{r} g_i((I-\mu J)^{-1}\varphi, f_i).$$

Choosing $\varphi(t) \equiv 1$ and taking into account (4.2) we obtain (4.7). Suppose now that (4.7) holds. Then we have, in view of $\mu J(I-\mu J)^{-1} = (I-\mu J)^{-1} - I$,

$$\begin{aligned}\mu JAe^{\mu t} &= ((I-\mu J)^{-1}-I)g_0 + \mu((I-\mu J)^{-1}-I) \sum_{i=1}^{r} g_i\hat{f}_i(\mu)\\ &= Ae^{\mu t} - g_0 - \mu \sum_{i=1}^{r} g_i\hat{f}_i(\mu).\end{aligned}$$

On the other side, there is

$$\mu AJe^{\mu t} = Ae^{\mu t} - g_0,$$

hence

$$(AJ - JA)e^{\mu t} = \sum_{i=1}^{r} g_i(e^{\mu t}, f_i).$$

Since $e^{\mu t}$ $(\mu \in \mathbb{C})$ is a complete system in L_p, this implies (4.4). ≡

PROOF OF THEOREM 4.1. We have to show that (4.6) is equivalent to (4.7). Let $\hat{a}(t,\mu)$ be defined by

$$\hat{a}(t,\mu) = \int_0^{\tau} a(t,s)e^{\mu s}ds.$$

Then $Ae^{\mu t} = \frac{d}{dt}\hat{a}(t,\mu) \in L_p$, hence $\hat{a}(t,\mu)$ depends absolutely continuous on t and $\hat{a}(t,\mu) = JAe^{\mu t} = 1 * Ae^{\mu t}$.

Therefore, using the relation $1 * (\delta + \mu e^{\mu t}) = e^{\mu t}$ and (4.1) one concludes from (4.7)

$$\hat{a}(t,\mu) = e^{\mu t} * g_0 + \mu \sum_{i=1}^{r}(e^{\mu t} * g_i)\hat{f}_i(\mu).$$

With the help of Lemma 4.1 we obtain that the function

$$\sum_{i=1}^{r} g_i(t-s) * f_i(s)$$

depends absolutely continuous on s and

$$a(t,s) = g_0(t-s) - \frac{\partial}{\partial s} \sum_{i=1}^{r} g_i(t-s) * f_i(s).$$

This yields (4.6a).

On the other side (4.6a) implies (4.7), applying all arguments in the reverse order. ≡

If A is invertible and has a reduction (4.4) then the inverse of A has a reduction

$$A^{-1}J - JA^{-1} = -\sum_{i=1}^{r}(\cdot, y_i)x_i.$$

Therefore, the following theorem holds.

THEOREM 4.2. Suppose $A \in \mathcal{T}_r^p$ is invertible. Then the inverse is given by

$$A^{-1}\varphi = \frac{d}{dt} \int_0^\tau c(t,s)\, \varphi(s)ds, \tag{4.8a}$$

$$c(t,s) = x_0(t-s) + \frac{\partial}{\partial s} \sum_{i=1}^{r} \int_0^{\min(t,s)} x_i(t-\sigma)\, y_i(s-\sigma)d\sigma, \tag{4.8b}$$

where x_i, y_i are the solutions of (4.5) and x_0 is the solution of $Ax_0 = 1$.

4.3. Sahnovich operators. As a standard example of JJ-Toeplitz-like operators, we consider in this subsection operators of the form

$$A\varphi = \frac{d}{dt} \int_0^\tau a(t-s)\, \varphi(s)ds \quad (0 \leq t \leq \tau), \tag{4.9}$$

where $a(t) \in L_p[-\tau,\tau]$ is such that the operator A is bounded in some L_p-space. In virtue of Lemma 3.2, (4.9) represents the general form of a "stationary" operator. Such operators are the continuous analogue of Toeplitz matrices. They will be referred to as Sahnovich operators.

Let us give some examples of Sahnovich operators. If $a(t) = \theta(t)-Jk(t)$, where $k(t) \in L_1[-\tau,\tau]$, then

$$A\varphi = \varphi(t) - \int_0^\tau k(t-s)\, \varphi(s)ds \quad (0 \leq t \leq \tau). \tag{4.10}$$

It is easily seen that this operator A is bounded in all L_p-spaces and, moreover, $\|A\|_p \leq \|k(t)\|_1 + 1$.

Another special class of Sahnovich operators are operators of the form

$$A\varphi = \sum_k \alpha_k\, \varphi(t - \delta_k), \tag{4.11}$$

where $\alpha_k \in \mathbb{C}$, $\sum |\alpha_k| < \infty$ and $\delta_k \in \mathbb{R}$, which are bounded in L_p and $\|A\|_p \leq \sum_k |\alpha_k|$. Operators (4.11) are obtained putting $a(t) = \sum_k \alpha_k\, \theta(t - \delta_k)$.

A third example is the operator of singular integration

$$S\varphi := \frac{1}{\pi i} \int_0^\tau \frac{\varphi(s)}{s-t}\, ds, \tag{4.12}$$

where the integral has to be understood in the sense of principal value. This operator is obtained putting $a(t) = \frac{i}{\pi} \log|t|$.

Now, let us show that Sahnovich operators belong to the class $\mathcal{T}_2^p$.

PROPOSITION 4.3. Suppose A is a bounded operator in L_p of the form (4.9), then

$$(AJ-JA)\varphi = (\varphi, a(-s))1 - (\varphi,1)\, a(t-\tau). \tag{4.13}$$

PROOF. We have

$$AJ\varphi = \int_0^\tau a(t-s)\,\varphi(s)ds - (\varphi,1)\,a(t-\tau)$$

and, on the other hand,

$$JA\varphi = \int_0^\tau a(t-s)\,\varphi(s)ds - (\varphi,a(-s)).$$

This yields (4.13). ≡

Let us still note the following relations

$$(AJ - J_\tau A)\varphi = (\varphi,a(\tau-s)) - (\varphi,1)\,a(t-\tau), \qquad (4.13')$$

where $J_\tau\varphi = -\int_t^\tau \varphi(s)ds$. That means, Sahnovich operators are JJ_τ-Toeplitz-like operators, too.

Next we determine the adjoint of a Sahnovich operator.

PROPOSITION 4.4. Suppose A is of the form (4.9), then A' is given by

$$A'\psi = -\frac{d}{dt}\int_t^\tau a(s-t)\,\psi(s)ds \qquad (0 \le t \le \tau). \qquad (4.14)$$

PROOF. From the proof of Lemma 4.2 we know that A' has the form

$$A'\psi = \frac{d}{dt}\int_0^\tau b(t,s)\,\psi(s)ds,$$

where $b(t,s) = A\theta_t(s) + h(s)$ and $h(s)$ is arbitrary. We have, by definition,

$$A\theta_t(s) = \frac{\partial}{\partial s}\int_0^t a(s,\sigma)d\sigma = a(s) - a(s-t).$$

This yields the representation (4.14). ≡

We note a further property, which is the continuous analogue of I, Prop. 0.2. To formulate this property we introduce the reflection operator $W\varphi := \varphi(\tau-t)$, which is the continuous analogue of the counter-identity.

PROPOSITION 4.5. Suppose A is of the form (4.9), then

$$WAW\,\varphi = -\frac{d}{dt}\int_0^\tau a(s-t)\,\varphi(s)ds. \qquad (4.15)$$

In particular, in L_2 there is $WAW = A'$.

The proof is an elementary computation. ≡

From Prop. 4.4 and 4.5 we obtain the following important fact.

COROLLARY 4.1. If A acts within L_p for $p < 2$, then the adjoint operator A' (acting within L_q, $q > 2$) can be extended by

$$\mathfrak{A}'\varphi := WAW\varphi$$

to a bounded operator $\mathfrak{A}'$ in L_p.

A further consequence of Prop. 4.4 and 4.5 is the following statement.

COROLLARY 4.2. A Sahnovich operator in L_2 which is Fredholm has vanishing index.

This assertion fails to be true in case $p \neq 2$. Indeed, the singular integral operator (4.12) has the index 1 if $p < 2$ and -1 if $p > 2$. We shall show in 6.7 that, provided $\dim \ker A < \infty$, the intersection of $\ker A$ and $\operatorname{im} J$ has a codimension at most one in $\ker A$. Hence, by Prop. 4.4 and 4.5, the dimensions of $\ker A$ and $\ker A'$ differ at most by one. Therefore, the following assertion holds true.

COROLLARY 4.3. For the index of a Fredholm Sahnovich operator A the following estimations are valid

$$0 \leq \operatorname{ind} A \leq 1 \quad \text{if } p \leq 2,$$

$$-1 \leq \operatorname{ind} A \leq 0 \quad \text{if } p \geq 2.$$

According to the general definition, the following equations are fundamental

$$Ax = 1 \quad , \quad Az = a(t-\tau), \tag{4.16a}$$

$$A'y = a(-t), \quad A'w = 1. \tag{4.16b}$$

In virtue of Prop. 4.4 and 4.5, we obtain that $w = Wx$ and $y = Wz$, provided that the equations (4.16) are solvable. Applying Theorem 4.2 we obtain the following assertion.

THEOREM 4.3. Suppose A is a Sahnovich operator in L_p, which is Fredholm. If the equations (4.16a) are solvable and $x, z \in L_p \cap L_q$ then A is invertible, and the inverse is given by (4.8a), where

$$c(t,s) = x(t-s) + \frac{\partial}{\partial s}\int_0^\tau (x(t-\sigma)z(\tau-s+\sigma) - z(t-\sigma)x(\tau-s+\sigma))d\sigma. \tag{4.17}$$

Now, let us present the third inversion variant (c.f. Section 1, Prop. 1.5) for Sahnovich operators. Instead of the equations (4.16a) we consider the following operator independent equations

$$Ax = 1, \quad Ax_1 = t \quad (0 \leq t \leq \tau). \tag{4.18}$$

PROPOSITION 4.6. Suppose the operator A is Fredholm, the equations (4.18) are solvable with $x, x_1 \in L_p \cap L_q$ and $(x,1) \neq 0$. Then A is invertible, and A^{-1} is given by (4.8a), (4.17), where

$$z = \frac{1}{(1,x)}(x_1 - Jx + (x, a(-t))x). \tag{4.19}$$

PROOF. Let z be defined by (4.19), then, due to (4.13),

$$Az = \frac{1}{(1,x)}(t - AJx + (x, a(-t))) = a(t-\tau).$$

It remains to apply Theorem 4.3. ≡

Now, let us discuss the specific of the symmetric case. The operator A is called symmetric if $A = \mathfrak{X}'(:= WAW)$, which is the same as $a(-t) = -a(t)$ (the case of a self-adjoint operator A in L_2 can be considered analogously). Instead of the two equations (4.16a) we consider only one equation

$$Au = \alpha + \beta\, a(\tau - t) \tag{4.20}$$

for fixed parameters $\alpha, \beta \neq 0$.

PROPOSITION 4.7. Suppose the operator A is Fredholm, and the equation (4.20) is solvable for some $\alpha, \beta \neq 0$. Then A is invertible, and A^{-1} is given by (4.8a), (4.17), where

$$x = \frac{1}{2\alpha}(u + Wu - \beta), \quad z = \frac{1}{2\beta}(-u + Wu - \beta). \tag{4.21}$$

PROOF. Let x, z be defined by (4.21). Then, in virtue of $WA = AW$ and

$$A1 = a(t) - a(t - \tau) = a(t) + a(\tau - t),$$

we obtain after an elementary calculation $Ax = 1$ and $Az = a(t-\tau)$. It remains to apply Theorem 4.3. ≡

We still consider the third inversion variant in the symmetric case. Instead of the two equations (4.18) we consider

$$Av = \gamma + t \tag{4.22}$$

PROPOSITION 4.8. Suppose the operator A is Fredholm, and the equation (4.22) is solvable for some $\gamma \neq -\frac{\tau}{2}$, where $(1,v) \neq 0$. Then A is invertible, and A^{-1} is given by (4.8a), (4.17), where x and z are defined via (4.19) and

$$x = \frac{1}{2\gamma + \tau}(v + Wv), \quad x_1 = \frac{1}{2\gamma+\tau}(\gamma(v - Wv) + \tau v).$$

The proof is a straightforward calculation. ≡

4.4. Further classes of JJ-Toeplitz-like operators.

The special cases of JJ-TLO presented below are continuous analogues of the special cases of matrices closed to Toeplitz considered in 2.1. First let us note that the sum and the product of JJ-TLO are such operators again. Next we deal with "paired" operators with displacement kernel

$$A\varphi = \frac{d}{dt}\int_0^{\tau_1} a(t-s)\,\varphi(s)ds + \frac{d}{dt}\int_{\tau_1}^{\tau} a_2(t-s)\,\varphi(s)ds.$$

In this case we have

$$(AJ - JA)\varphi = (\varphi, \chi\tau_1)\, g(t) - (\varphi, 1)\, a_2(t-\tau) + (\varphi, f)1,$$

where

$$\chi\tau_1(t) = \begin{cases} 1 : 0 \le t \le \tau_1 \\ 0 : \tau_1 < t \le \tau \end{cases}, \quad g(t) = a_2(t-\tau_1) - a_1(t-\tau_1), \quad f(t) = \begin{cases} a_1(-t) : 0 \le t \le \tau_1 \\ a_2(-t) : \tau_1 < t \le \tau. \end{cases}$$

Hence $A \in \mathcal{J}_3^p$, provided that A is bounded in L_p. In generalization of paired operators, one may consider operators (4.3) with "mosaic kernel"

$$a(t,s) = a_{ik}(t-s) \; (\tau_{i-1} < t \le \tau_i;\; \sigma_{k-1} < s \le \sigma_k;\; i=1,\dots,m;\; k=1,\dots,n;\; \tau_0 = \sigma_0 = 0;\; \tau_m = \sigma_n = \tau).$$

As is easily seen, such operators belong to $\mathcal{T}^p_{m+n}$.
Finally, we consider bounded operators of the form $A = C + B$, where

$$C\varphi = \frac{d}{dt}\int_0^\tau c(t-s)\,\varphi(s)ds, \quad B\varphi = \frac{d}{dt}\int_0^\tau b(t+s)\,\varphi(s)ds \quad (0 \le t \le \tau).$$

Analogously to Toeplitz plus Hankel matrices, the consideration of such operators can be replaced by that one of the operator
$\mathfrak{X}\varphi = \frac{d}{dt}\int_{-\tau}^{\tau} a(t,s)\,\varphi(s)ds$ in $L_p(-\tau,\tau)$, where $a(t,s)$ is of mosaic type

$$a(t,s) = \begin{bmatrix} c(s-t) & b(s-t) \\ b(t-s) & c(t-s) \end{bmatrix}.$$

Analogously to Prop. 2.3 the following assertion can be proved.

PROPOSITION 4.9. The operator $\mathfrak{X}$ is invertible if and only if $A = C + B$ and $C - B$ are invertible. If $\mathfrak{X}$ is invertible, then the inverse of A is given by $A^{-1} = P\,\mathfrak{X}^{-1} R$, where $P : L_p(-\tau,\tau) \to L_p$ is the restriction, $(P\varphi)(t) = \varphi(t)$ $(t \in [0,\tau])$, and $R : L_p \to L_p(-\tau,\tau)$ is defined by $(R\varphi)(t) = \varphi(|t|)$.

4.5. Generating functions. We saw in Part I that the generating function of a matrix is a useful instrument for representing inversion formulae. This concept can be also introduced for operators in L_p.
Let C be a bounded operator in L_p. Then we define

$$C(\lambda,\mu) := (Ce^{\mu t}, e^{\lambda t}).$$

The function $C(\lambda,\mu)$ $(\lambda,\mu \in \mathbb{C})$ will be referred to as generating function of C. Clearly, C is uniquely defined by $C(\lambda,\mu)$. The following proposition is the continuous analogue of I, Prop. 1.9.

PROPOSITION 4.10. The generating function of C and D_C are connected via the formula

$$C(\lambda,\mu) = \frac{1}{\lambda+\mu}(\mu\lambda D_C(\lambda,\mu) + \lambda p(\lambda) + \mu q(\mu)e^{\lambda\tau}),$$

where $p(\lambda) = (C1, e^{\lambda t})$ and $q(\mu) = (Ce^{\mu t}, 1)$.

The proof is an elementary computation. ≡
As an application of Prop. 4.8, let us determine the generating function of the inverse of a Sahnovich operator.

PROPOSITION 4.11. Let A be an invertible operator (4.9), then $A^{-1} = C$ has the generating function

$$C(\lambda,\mu) = \frac{1}{\lambda+\mu}(a(\lambda)b(-\mu)e^{\mu\tau} - b(\lambda)a(-\mu)e^{\mu\tau}) \qquad (4.23)$$

where $a(\lambda) = \lambda\hat{x}(\lambda)$ and $b(\lambda) = \lambda\hat{z}(\lambda) + e^{\lambda\tau}$.

The proof is a straightforward calculation. ≡

We obtained that the inverse of Sahnovich operators is the continuous analogue of a T-Bezoutian. Therefore, it seems to be natural to consider operators with generating functions (4.23) separately. It is easily seen that such a operator is bounded if the function $a(i\omega)$ and $b(i\omega)$ are bounded on $\mathbb{R}$. These arguments lead to a sufficient invertibility criterion without the Fredholmness assumption.

REMARK 4.2. Suppose the equations (4.16a) are solvable and $\omega\hat{x}(i\omega)$, $\omega\hat{z}(i\omega)$ are bounded on $\mathbb{R}$. Then A is invertible.

4.6. DD-Toeplitz-like operators. Let D denote the operator of differentiation in L_p and W_p its domain. In this subsection we discuss the class of DD-Toeplitz-like operators. Because D is unbounded, this concept is not covered by the general definition of a TLO, and we have to define it separately.
The operator A in L_p will be called DD-TLO if $AW_p \subseteq W_p$, $A'W_q \subseteq W_q$ and $r_A := \operatorname{rank}(AD - DA) < \infty$. The class of such operators with $r_A \leq r$ will be denoted by $\mathcal{D}^p_r$. It is easily seen that $\mathcal{D}^p_r$ is a subclass of $\mathcal{T}^p_{r+1}$ and, conversely, an operator $A \in \mathcal{T}^p_r$ satisfying $AW_p \subseteq W_p$, $A'W_q \subseteq W_q$ belongs to $\mathcal{D}^p_{r+1}$. If A is an operator satisfying the latter conditions, then $AJ - JA = \sum_{i=1}^{r}(\cdot, f_i)g_i$ for some $g_i \in W_p$ and $f_i \in W_q$. Applying Theorem 4.1 one obtains that A is an integral operator $A\varphi = \int_0^\tau K(t,s)\,\varphi(s)ds$, where

$$K(t,s) = \alpha\delta(t-s) - h(t-s) - \sum_{i=1}^{r} \int_0^\tau g_i'(t-\sigma)\, f_i'(s-\sigma)d\sigma,$$

$$\alpha = g_0(0) - \sum_{i=1}^{r} f_i(0)g_i(0), \quad h(t) = g_0'(t) - \sum_{i=1}^{r} f_i(0)g_i'(t).$$

Therefore, the following assertion is true.

PROPOSITION 4.12. The operators $A \in \mathcal{D}^p_r$ have the form $A = \alpha I - K$, where K is compact. In particular, A is Fredholm if and only if $\alpha \neq 0$, and $\alpha \neq 0$ implies $\operatorname{ind} A = 0$.

For constructing inversion formulae for DD-TLO one can use an approach different to that one presented above, which bases upon the following lemma.

LEMMA 4.4. Suppose $A \in \mathcal{D}^p_r$ and $D_A := AD - DA$, then

$$e^{-\sigma D} A - Ae^{-\sigma D} = \int_0^\sigma e^{(s-\sigma)D}\, D_A\, e^{-sD} ds.$$

The proof is straightforward verification. ≡

With the help of this formula one can construct, for example, inversion formulae for operators

$$A\varphi = \varphi(t) - \int_0^\tau k(t-s)\,\varphi(s)ds \quad (0 \leq t \leq \tau), \tag{4.24}$$

where $k(t) \in L_1(-\tau,\tau)$. Such operators belong to $\mathcal{D}^p_1$ and

$$(AD - DA)\varphi = \varphi(0)k(t) - \varphi(\tau)k(t-\tau).$$

In this way the following proposition can be proved.[1]

PROPOSITION 4.13. Suppose the equations

$$Au = k(t), \quad Av = k(t-\tau) \tag{4.25}$$

are solvable. Then A is invertible in all spaces L_p, and the inverse is given by

$$A^{-1}\psi = \psi(t) + \int_0^\tau \gamma(t,s)\,\psi(s)ds,$$

$$\gamma(t,s) = u(t-s)+v(t-s) + \int_0^\tau (u(t-\sigma)v(\tau+\sigma-s)-v(t-\sigma)u(\sigma-s+\tau))d\sigma.$$

We finally remark that the third inversion variant can be applied to operators (4.24) assuming $k(t) \in W_1(-\tau,\tau)$ and replacing (4.25) by

$$Au = k(t) \qquad Aw = k'(t). \tag{4.26}$$

The systems (4.25) and (4.26) are connected via the formula

$$u(\tau)v(t) = w(t) - u'(t) + u(t)u(0).$$

Therefore, the inverse operator can be constructed with the help of $u(t)$ and $w(t)$, provided that $x(\tau) \neq 0$.

4.7. Non-Fredholm operators. Unfortunately the integral operators occurring in application are often not Fredholm in L_p but, on the other side, the fundamental equations are solvable. In such cases one tries to construct right inverse operators acting in a pair of spaces. In this subsection we give some idea of such an approach not going into details.

Let W_p^m denote the Sobolev space of all functions $\varphi(t)$ on $[0,\tau]$, for which $D^k\varphi \in L_p$ $(k = 0,\dots,m)$, and $\overset{\circ}{W}{}_p^m$ the subspace of W_p^m containing all functions $\varphi(t)$ with $\varphi(0) = \varphi(\tau) = 0$.

THEOREM 4.4. Suppose $A \in \mathcal{T}^p_r$, and the equations (4.5) and $Ax_0 = 1$ are solvable. Then $\operatorname{im} A \supseteq W_p^2$, and the operator $C : \overset{\circ}{W}{}_p^2 \to L_p$ defined by

$$C\varphi = \int_0^\tau c(t,s)\,\varphi''(s)ds, \tag{4.27a}$$

$$c(t,s) = x_0(t-s) + \sum_{i=1}^r \int_0^{\min(t,s)} x_i(t-\sigma)y_i(s-\sigma)d\sigma, \tag{4.27b}$$

is right inverse to A.

PROOF. We have, by assumption, $1 \in \operatorname{im} A$. This implies $t \in \operatorname{im} A$. .

---

[1] For other proofs see Notes and Comments.

Therefore, to prove the inclusion $\operatorname{im} A \supseteq W_p^2$ we have to show $\operatorname{im} A \supseteq \mathring{W}_p^2$. We assume $\psi \in \mathring{W}_p^2$. If $x(\mu)$ denotes the solution of

$$Ax(\mu) = e^{\mu t}, \tag{4.28}$$

which exists and is given by (1.18) for all $\mu \in \mathbb{C}$, then a standard calculation shows that the integral

$$\varphi = \frac{1}{2\pi} \int_{-\infty}^{\infty} x(i\omega)\, \psi(i\omega) d\omega$$

exists and, therefore, represents a right inverse operator of A. Furthermore, after some transformations (4.27) is obtained. ≡

In the remainder of this subsection <u>we shall assume $p < 2$</u> and, therefore, $q > 2$ $(\frac{1}{p} + \frac{1}{q} = 1)$. Not rarely one has the situation that the fundamental equations (4.5b) are not solvable in L_q, but replacing A' by its extension $\mathfrak{A}'$ to L_p they are, provided such an extension exists. The simplest example of such an operator is the operator S of singular integration, since $\mathfrak{S}' = S$, S is an epimorphism but no monomorphism in L_p (in virtue of $(t(\tau-t))^{-1/2} \in \ker S$).
We show now how in this case the solutions of (4.28) can be constructed. Suppose $A \in \mathcal{T}_r^p$ has the reduction (4.4). We consider the fundamental equations

$$Ax_i = g_i, \quad Ax_o = 1, \quad \mathfrak{A}' y_i = f_i \quad (x_o, x_i, y_i \in L_p;\ i=1,\dots,r). \tag{4.29}$$

For simplicity we introduce the operator $X : \mathbb{C}^r \to L_p$, $X(\alpha_i)_1^r = \sum_{i=1}^r \alpha_i x_i$, and the analogously defined operators $Y : \mathbb{C}^r \to L_p$, $F, G : \mathbb{C}^r \to L_q$. In particular, $D_A = GF'$. We introduce, furthermore, the matrix $\Omega := Y'G - F'X$ and the vector $p = F'x_o - Y'1$. Let us note that in case A is invertible we have $\Omega = 0$ and $p = 0$. On the other hand, the operator S is an example showing that $\Omega \neq 0$ is possible.

<u>PROPOSITION 4.14.</u> Suppose the equations (4.29) are solvable and μ is such that the matrix $I - \mu\Omega$ is regular. Then

$$x(\mu) = (I-\mu J)^{-1}(x_o - \mu X(I-\mu\Omega)^{-1}(Y' e^{\mu t} + p)) \tag{4.30}$$

is a solution of (4.28).

<u>PROOF.</u> Let $x(\mu)$ be a solution of

$$(I - \mu(J - XF'))x(\mu) = x_o, \tag{4.31}$$

then $x(\mu)$ satisfies (4.28), too. It remains to show that (4.30) is a solution of (4.31). This equation is solvable for all μ except a countable set. Assuming that (4.31) is solvable we obtain

$$x(\mu) = (I - \mu J)^{-1}(x_o - \mu X\xi(\mu)),$$

where $\xi(\mu) = F'x(\mu)$. Since the range of $(I - \mu J)^{-1} - I$ is contained in W_p and, therefore, in L_q, we have $F'((I-\mu J)^{-1}-I) = Y'A((I-\mu J)^{-1}-I)$.

We conclude using this argument

$$\xi(\mu) = Y'Ax(\mu) - Y'A(x_0 - \mu X\xi(\mu)) + F'(x_0 - \mu X\xi(\mu))$$

$$= Y'e^{\mu t} + p + \mu\Omega\,\xi(\mu).$$

This yields (4.30). To obtain (4.30) for all μ we use arguments of continuity. ≡

Now, similar to Theorem 4.1, a right inverse operator $C : W_p^2 \to L_p$ of A can be constructed. It is remarkable that in case Ω is regular we have even $W_p \subseteq \operatorname{im} A$, because here $\|(I - \mu\Omega)^{-1}\| \xrightarrow[\mu\to\infty]{} 0$.

4.8. Integral equations on the half-line. Let us shortly discuss how the considerations above have to be modified in case $\tau = \infty$. In particular, the question is how to apply the TLO approach to bounded operators of the form

$$A\varphi = \frac{d}{dt}\int_0^\infty a(t-s)\,\varphi(s)ds \quad (0 \le t < \infty). \tag{4.32}$$

Special cases of such operators are, for example, Wiener-Hopf integral operators, the singular integral operator on the half-line and integro-difference operators in the sense of I. GOHBERG, I. FELDMAN [1]. The integration operator is not suitable to obtain a reduction, because it is not bounded in L_p. For that reason it seems to be more convenient to take $U = V = J(\beta)$, where

$$J(\beta)\varphi = \int_0^t e^{\beta(t-s)}\,\varphi(s)ds \quad (0 \le t < \infty)$$

for $\operatorname{Re}\beta < 0$, since these operators, in view of $e^{-\beta t} \in L_1(0,\infty)$, are bounded. We obtain the following reduction for operators of the form (4.23)

$$AJ(\beta) - J(\beta)A = (\cdot, f_\beta(t))e^{\beta t},$$

where

$$f_\beta(t) = -\frac{d}{dt}\int_{-\infty}^0 a(s-t)e^{-\beta s}ds \ (=a(-t) + \beta\int_t^\infty e^{-\beta(t-\sigma)}a(-\sigma)d\sigma).$$

We finally note that Wiener-Hopf integral operators

$$A\varphi = \varphi(t) - \int_0^\infty k(t-s)\,\varphi(s)ds \quad (0 \le t < \infty), \tag{4.33}$$

where $k(t) \in L_1(\mathbb{R}) \cap L_p(\mathbb{R})$ can be also regarded as DD-TLO. In fact, there is

$$(AD - DA)\varphi = -\varphi(0)k(t), \tag{4.34}$$

provided that φ is absolutely continuous. Basing upon (4.34) one can construct inversion formulae, in particular, one obtains the formulae well-known from the classical Wiener-Hopf theory (see M.G. KREIN [1]).

5. SINGULAR INTEGRAL AND TOEPLITZ OPERATORS

The main aim of this section is to show that one-dimensional singular-integral and Toeplitz operators are special types of TLO when choosing U and V as the operator of multiplication by the independent variable.

<u>5.1. The algebra of singular integral operators.</u> Throughout this section Γ will denote a rectifiable oriented (closed or open) Liapunov contour in the complex plane. We consider the operator of singular integration along Γ

$$(S_\Gamma\varphi)(t) = \frac{1}{\pi i} \int_\Gamma \frac{\varphi(\tau)}{\tau-t}\, d\tau \qquad (t \in \Gamma). \tag{5.1}$$

The integral has to be understood in the sense of principal value. It is well-known that by (5.1) a linear bounded operator is defined in the space $L_p = L_p(\Gamma)$ of p-summable functions on Γ if $1 < p < \infty$, and furthermore, S_Γ is not bounded in $L_1 := L_1(\Gamma)$ and $L_\infty := L_\infty(\Gamma)$. Let $\tilde{L}_\infty$ denote the intersection of all spaces L_p $(p < \infty)$. Then S_Γ can be regarded as an operator $S_\Gamma : L_\infty \to \tilde{L}_\infty$. The operator (5.1) can also be regarded in the space L_p^n of n component vector function $(\varphi_i(t))_1^n$ with $\varphi_i(t) \in L_p$.

In all what follows for an arbitrary set M the set of column vectors with components from M will be denoted by M^n and the set of nxn matrices with entries from M by $M^{n\times n}$. Further, we introduce the following notation

$$(\varphi,\psi) = \frac{1}{\pi i} \int_\Gamma \psi(t)^T \varphi(t)dt.$$

This notation will be used in case $\varphi \in L_p^{n\times n}$ or $\varphi \in L_p^n$ and $\psi \in L_q^{n\times n}$ or $\psi \in L_q^n$ $(\frac{1}{p} + \frac{1}{q} = 1)$, respectively. Let A be a linear bounded operator in L_p^n. Then by

$$(A\varphi,\psi) = (\varphi,A^T\psi) \quad (\varphi \in L_p^n, \quad \psi \in L_q^n)$$

an operator A^T in L_q is defined, which is called <u>transposed to A</u>.[1)] Obviously, there is $S_\Gamma^T = - S_\Gamma$.

Let $\mathfrak{A}$ be an algebra of essential bounded functions on Γ containing the functions $\varphi_k(t) = t^k$ for $k = -1,0,1$. To any matrix function $a(t) \in \mathfrak{A}^{n\times n}$ corresponds the operator of multiplication $(a\varphi)(t)=a(t)\varphi(t)$ in L_p^n. The (not closed) algebra of operators in L_p^n generated by S_Γ and the multiplication operators generated by $a(t) \in \mathfrak{A}^{n\times n}$ will be denoted by $\mathfrak{R}_p^n(\mathfrak{A})$. In particular, we set $\mathfrak{R}_p^n := \mathfrak{R}_p^n(L_\infty)$. The elements

1) We remark that A^T does not coincide with A'. The operator A' is usual defined as follows $[A\varphi,\psi] = [\varphi,A'\psi]$, where $[\varphi,\psi] = \int_\Gamma \psi(t)^T \varphi(t)|dt|$.

of $\mathfrak{R}_p^n$ will be called singular integral operators.

Let U_Γ denote the operator of multiplication by the independent variable in L_p^n, $U_\Gamma\varphi = t\varphi(t)$. Clearly, U_Γ commutes with all multiplication operators. Furthermore, we have

$$(S_\Gamma U_\Gamma - U_\Gamma S_\Gamma)\varphi = \sum_{k=1}^{n}(\varphi, e_k)e_k. \tag{5.2}$$

This leads to the following.

THEOREM 5.1. A linear bounded operator in L_p^n is a $U_\Gamma U_\Gamma$-Toeplitz-like operator in all spaces L_p^n $(p > 1)$ if and only if it is a singular integral operator. Moreover, any singular integral operator A admits a representation

$$A\varphi = a(t)\varphi(t) + \sum_{j=1}^{r} g_j(t)\ S_\Gamma\ f_j(t)^T\ \varphi(t), \tag{5.3}$$

where $g_j(t)$, $f_j(t)$ and the columns of $a(t)$ belong to $\mathfrak{L}_\infty$.

For the proof we need the following assertion.

LEMMA 5.1. Suppose C is a bounded operator mapping from L_p^n $(p > 1)$ into $L_{p-\varepsilon}^n$ for all $0 < \varepsilon < p-1$, and C commutes with U_Γ. Then C is the operator of multiplication by a matrix function $a(t)\in\mathfrak{L}_\infty^{n\times n}$. If C is, moreover, bounded in L_p^n then $c(t)$ is essential bounded.

PROOF. The relation $CU_\Gamma = U_\Gamma C$ implies

$$Ct^k e_j = t^k Ce_j \in L_{p-\varepsilon}^n.$$

Denoting the matrix with columns Ce_j by $a(t)$ we obtain

$$C\varphi(t) = a(t)\ \varphi(t) \tag{5.4}$$

for all vector polynomials $\varphi(t)$. Since the vector polynomials are dense in L_p^n, we conclude using classical arguments that (5.4) holds for all $\varphi \in L_p^n$ and $c(t) \in L_q^{n\times n}$. $\equiv$

PROOF OF THEOREM 5.1. From (5.2) and the properties of TLO we conclude that any singular integral operator A is a $U_\Gamma U_\Gamma$-TLO and

$$D_A := AU_\Gamma - U_\Gamma A = \sum_{j=1}^{r}(\cdot, f_j)g_j, \tag{5.5}$$

for some g_i, $f_j \in \mathfrak{L}_\infty^n$. On the other side, let A be a $U_\Gamma U_\Gamma$-TLO in all spaces L_p^n. Then, by definition, a representation (5.5) holds. We consider the operator

$$\mathfrak{X}\varphi = \sum_{j=1}^{r} g_j(t)\ S_\Gamma\ f_j(t)^T\ \varphi(t)$$

acting from L_p^n into $L_{p-\varepsilon}^n$ for all ε with $0 < \varepsilon < p-1$. Furthermore, we have $D_A = D_{\mathfrak{X}}$. Therefore, the operator $C = A - \mathfrak{X}$ commutes with U_Γ. Applying Lemma 5.1 we obtain that C is a multiplication operators by a matrix function $a(t) \in \tilde{L}_\infty^{n\times n}$. Hence, the representation (5.3) holds. This completes the proof. ≡

5.2. Inversion of singular integral operators. Let A be a singular integral operator in L_p^n. Then, according to Theorem 5.1, A has the form (5.3), and its $U_\Gamma U_\Gamma$-reduction D_A has the form (5.5). Thus, fundamental equations corresponding to A are

$$Ax_i = g_i \qquad (x_i \in L_p^n) \tag{5.6a}$$

and

$$A^T y_i = f_i \qquad (y_i \in L_q^n). \tag{5.6b}$$

The transposed A^T of A is given by

$$A^T\psi = a(t)^T\psi(t) - \sum_{i=1}^{r} f_j(t)\, S_\Gamma\, g_j(t)^T\, \psi(t).$$

PROPOSITION 5.1. The reduction D_A is complete (i.e. (1.1) is satisfied) if and only if A is Fredholm.

PROOF. If the reduction is complete, then A is Fredholm, by Prop. 1.2.
On the other hand, let A be Fredholm and $\varphi \in \ker \mathrm{Obs}_k A$. Then $t^i\varphi \in \ker A$ for $i = 0,\dots,k$. If $k > \dim\ker A$ this implies $p(t)\,\varphi(t) \equiv 0$ for some non-trivial polynomial $p(t)$, which yields $\varphi = 0$. Therefore, $\ker \mathrm{Obs}_k A = 0$. That means, the reduction is complete. ≡

An immediate consequence of Theorem 1.1 is the following.

PROPOSITION 5.2. Suppose A is Fredholm in L_p^n and the equations (5.6) are solvable. Then A is invertible in L_p^n.

To obtain an inversion formula we consider besides the equations (5.6a), (5.6b) the following equations

$$Ax_k^o = e_k \qquad (k = 0,\dots,n-1). \tag{5.6c}$$

We denote by $x_o(t)$ the matrix function with columns $x_k^o(t)$.

THEOREM 5.2. Suppose $A \in \mathfrak{K}_p^n$ is given by (5.3), and the fundamental equations (5.6) are solvable for all p with $1 < p < \infty$. Then the inverse of A is given by

$$A^{-1}\psi = z(t)\psi(t) - \sum_{i=1}^{r} x_j(t)\, S_\Gamma\, y_j(t)^T\, \psi(t), \tag{5.7}$$

where

$$z(t) = x_o(t) + \sum_{j=1}^{r} x_j(t)\, S_\Gamma\, y_j(t)^T.\ ^{1)} \tag{5.8}$$

PROOF. By Prop. 5.2, A is invertible in all spaces L_p^n. Furthermore,

---

1) The application of S_Γ to row vectors is also defined by (5.1).

we have

$$D_{A^{-1}} = - \sum_{j=1}^{r} (\cdot, y_j) x_j .$$

In virtue of Lemma 5.1, this implies that A^{-1} has the form (5.7) for some matrix function $z(t)$. The equality (5.8) results applying (5.7) to the constant functions e_k. ≡

Let us discuss now the special case of singular integral operators from $\mathfrak{R}_p^n(\Lambda_\alpha)$, $0 < \alpha < 1$, where Λ_α denotes the space of Hölder continuous functions on a closed contour Γ with the exponent α. Since the operator S_Γ acts within Λ_α, any operator $A \in \mathfrak{R}_p^n(\Lambda_\alpha)$ can be represented in the form (5.3) with $a(t) \in \Lambda_\alpha^{n\times n}$, $g_j(t)$, $f_j(t) \in \Lambda_\alpha^n$. The following result is well-known:

The operator (5.3) with continuous coefficients is Fredholm if and only if

$$\det(a(t) \pm \sum_{j=1}^{r} g_j(t)\, f_j(t)^T) \neq 0 \quad (t \in \Gamma). \tag{5.9}$$

If the latter condition is satisfied, then $cI + dS_\Gamma$ with

$$c(t) \pm d(t) = (a(t) \pm \sum_{j=1}^{r} g_j(t)\, f_j(t)^T)^{-1} \tag{5.10}$$

is a regularizator of A.

<u>THEOREM 5.3.</u> Suppose A is given by (5.3), where $a(t) \in \Lambda_\alpha^{n\times n}$, $g_j(t)$, $f_j(t) \in \Lambda_\alpha^n$, and (5.9) is satisfied. If (5.6a), (5.6b) are solvable in Λ_α^m then A is invertible in L_p, and A^{-1} is given by (5.7) with

$$z(t) = c(t). \tag{5.11}$$

<u>PROOF.</u> It remains to prove (5.11). Since $cI + dS_\Gamma$ is a regularizator of A, it differs from A^{-1} at least by a compact operator. On the other side, (5.7) can be written in the form $A^{-1} = zI + wS_\Gamma + T$, where T is compact. Because an operator $aI + bS_\Gamma$ is compact only in case $a = b = 0$ this yields (5.11). ≡

<u>5.3. Inversion of Toeplitz operators.</u> In this subsection the contour Γ will be the unit circle $\mathbb{T}$ and $S := S_{\mathbb{T}}$. It is well-known that $S^2 = I$ in L_p $(1 < p < \infty)$. Therefore, the operators $P := \frac{1}{2}(I+S)$ and $Q := \frac{1}{2}(I-S)$ are complementary projections in L_p. As usual, we denote $H_p := PL_p$. To any essential bounded measurable matrix function we associate the operator T_a acting within H_p^n defined by

$$T_a\varphi = Pa(t)\varphi(t) \quad (t \in \mathbb{T}). \tag{5.12}$$

Operators of this form are called <u>Toeplitz operators.</u>

Toeplitz operators admit a discrete interpretation as discrete Wiener-Hopf operators, i.e. as operators generated by an infinite Toeplitz matrix. In fact, let h_p^n denote the space of all sequences $(\xi_k)_0^\infty = F\varphi$,

$\varphi \in H_p^n$, where F denotes the Fourier transformation $F\varphi = \frac{1}{2}((\varphi, t^{-k-1}))_0^\infty$. Then FT_aF^{-1} is just the operator in h_p^n generated by the infinite block Toeplitz matrix $T(a) = [a_{j-k}]_0^\infty$, where a_j $(j \in Z)$ denote the (matrix) Fourier coefficients of $a(t)$.
Toeplitz operators are closed to singular integral operators. This shows the relation

$$(PaP+Q)(I+QaP) = \tfrac{1}{2}(a+I_n)I + \tfrac{1}{2}(a-I_n)S$$

where $(I+QaP)^{-1} = I-QaP$. But Toeplitz operators can also be regarded as UU-TLO, where U is the multiplication by tI_n or, in the discrete interpretation, the forward shift in h_p^n. Indeed, we have

$$(T_aU - UT_a)\varphi = (\varphi, a), \tag{5.13}$$

which is

$$T(a)U - UT(a) = e_0 f^T, \tag{5.13'}$$

where e_0 is the block column $e_0 = (I_n 0 \ldots 0)$ and f^T the block row $f^T = [a_{-1}\ a_{-2} \ldots]$. The fundamental equations corresponding to $T(a)$ can be written as two block vector equations

$$T(a)x = e_0 \qquad x = (x_k)_0^\infty \in h_p^{n\times n} \tag{5.14a}$$

$$T(a)^T z = f \qquad z = (z_k)_0^\infty \in h_p^{n\times n}. \tag{5.14b}$$

For Toeplitz operators it is convenient to apply the second inversion variant, which consists in replacing (5.14b) by

$$T(a)^T y = e_0 \qquad y = (y_k)_0^\infty \in h_p^{n\times n}. \tag{5.14c}$$

The connection between z and x is given by

$$zy_0 = -U^T y.$$

It is easily seen that y_0 is regular, provided that $T(a)$ is invertible. Conversely, the solvability of (5.14a), (5.14c) and the regularity of y_0 imply the invertibility of $T(a)$, according to the results of Section 1. Applying I, Lemma 1.1 we obtain the following classical result of the Wiener-Hopf theory.

<u>PROPOSITION 5.3.</u> Suppose the operator $T(a)$ is Fredholm. Then $T(a)$ is invertible if and only if (5.14a), (5.14c) are solvable and x_0 is regular. If the latter conditions are satisfied the inverse operator is generated by the matrix

$$T(a)^{-1} = \begin{bmatrix} x_0 & & & \\ x_1 & x_0 & & \\ x_2 & x_1 & x_0 & \\ \vdots & \ddots & \ddots & \ddots \end{bmatrix} y_0^{-1} \begin{bmatrix} y_0 & y_1 & y_2 & \cdots \\ & y_0 & y_1 & \ddots \\ & & y_0 & \ddots \\ & & & \ddots \end{bmatrix} \tag{5.15}$$

In the continuous interpretation T_a of $T(a)$ it is convenient to identify the dual space of H_p^n with QL_p^n via the bilinear form $(\cdot,\cdot)$. The fundamental equations are given by

$$T_a x = I_n, \quad T_a' z = Qa, \quad T_a' y = t^{-1} I_n \quad (x \in H_p^n; \ y,z \in QL_p^n); \tag{5.16}$$

the inverse of T_a is constructed via the formula

$$T_a^{-1}\varphi = x(t)\ y_0^{-1}\ Py(t)\varphi(t). \tag{5.17}$$

We still note that, clearly, a sum of products of Toeplitz operators is a UU-TLO. Therefore, inversion formulae for such operators can be constructed. Using the idea of (finite) Toeplitz plus Hankel matrix inversion one can, furthermore, construct inversion formulae for Toeplitz plus Hankel operators in h_p^n (H_p^n). We finally remark that, of course, Toeplitz operators and their generalizations can be considered in other Banach spaces, for example in the space l_p^n $(p \geq 1)$.

6. KERNEL STRUCTURE AND PARTIAL INDICES OF TLO

We proceed in this section the investigation of abstract Toeplitz-like operators (TLO). All notations and definitions are taken from Section 1. The main goal of this section is the generalization of the results about the kernel structure of Toeplitz matrices (c.f. I, Section 5). In that way some results of the theory of matrix-valued Toeplitz operators (the "Wiener-Hopf theory") are generalized, too. We shall define the concept of partial indices originated in the Wiener-Hopf theory for general TLO. Considerable attention is dedicated to the relationship between the kernel structure of TLO and the solvability of the fundamental equations. The corresponding results can be applied to obtain new proofs and statements concerning the problem of Wiener-Hopf factorization of matrix functions.

<u>6.1. Kernel structure theorem.</u> We consider in this section UV-Toeplitz-like operators, where $U \in \mathcal{L}(\mathbb{E})$ and $V \in \mathcal{L}(\mathbb{F})$ are fixed (reduction) operators; N is a fixed finite-dimensional (normalization) operator. For simplicity we make the following additional assumptions to the TLO under consideration. We remark that some of the forthcoming statements are also true without some of these assumptions.

<u>DEFINITION 6.1.</u> We denote by $\mathcal{T}_r^\circ$ the class of all operators $A \in \mathcal{L}(\mathbb{E},\mathbb{F})$ satisfying the following conditions:

(1) $\operatorname{rank}(D_A + N) = r$

(2) The completeness conditions (C1), (C2) (see Subsection 1.1) are fulfilled.

(3) $\operatorname{im}(D_A + N) = \operatorname{im} D_A + \operatorname{im} N$, $\operatorname{im}(D_A + N)' = \operatorname{im} D_A' + \operatorname{im} N'$.

Clearly, in the case of $N = 0$ the condition (3) can be omitted and (2) results in (1.1). We remark, furthermore, that Toeplitz and Hankel matrices (as well as $d\times d$ block Toeplitz and Hankel matrices) satisfy Definition 6.1, where $r = 2$ $(r = 2d)$ defining U, V and N as in Section 1. Finally, we note that (3) is equivalent to the conditions

$$(3') \quad \ker(D_A+N) = \ker D_A \cap \ker N, \quad \ker(D_A+N)' = \ker D_A' \cap \ker N'$$

and that the operators of the class $\mathcal{T}_r^o$ are Fredholm, by Prop. 1.2.

We are going to generalize the theorem about the kernel structure of Hankel matrices. For this purposes we have to generalize the concept of a U-chain introduced in I, Section 5.

<u>DEFINITION 6.2.</u> A sequence of vectors $x_o,\dots,x_{\nu-1} \in \mathbb{E}$ will be called <u>(U,N)-chain of length ν</u> if

$$\text{(a)} \quad x_{k+1} = Ux_k \qquad \text{(b)} \quad Nx_k = 0 \quad (k = 0,\dots,\nu-2).$$

In this sense the U-chains considered in Part I are, precisely speaking, (U_n,N)-chains, where U_n is the forward shift in $\mathbb{C}^n$ and $N = e_{m-1}e_{n-1}^T \in \mathbb{C}^{m\times n}$. In general, we ask about bases consisting of (U,N)-chains spanning the kernel of an operator $A \in \mathcal{T}_r$ and about bases of (V',N')-chains spanning $\ker A'$. We formulate now the main result of this subsection.

<u>THEOREM 6.1.</u> Suppose $A \in \mathcal{T}_r^o$, then there is a basis of $\ker A$ consisting of $r_o \le r$ (U,N)-chains. For the minimal value of r_o the lengths of the chains are uniquely determined by A.

For proving the theorem we have to investigate the structure of the subspaces $\mathcal{A}_k$ defined by

$$\mathcal{A}_k := \ker \begin{bmatrix} \mathrm{Obs}_k A \\ \mathrm{Obs}_{k-1} N \end{bmatrix} = \ker A \cap \ker \begin{bmatrix} \mathrm{Obs}_{k-1} D_A \\ \mathrm{Obs}_{k-1} N \end{bmatrix}$$

more in detail. First we observe that $\ker A = \mathcal{A}_o \supseteq \mathcal{A}_1 \supseteq \dots \supseteq \mathcal{A}_{q+1} = \{0\}$. Let q denote the greatest index k, for which $\mathcal{A}_k \neq \{0\}$. Such an index exists in view of condition (2). The following proposition is an obvious consequence of the definitions above and the normal solvability of the operator $D_A + N$.

<u>PROPOSITION 6.1.</u> Suppose $x \in \ker A$, then the following conditions are equivalent:

1) $x \in \mathcal{A}_k$.
2) $x, Ux, \dots, U^{k-1}x$ is a (U,N)-chain belonging to $\ker A$.
3) $D_A x = \dots = D_A U^{k-1}x = Nx = \dots = NU^{k-1}x = 0$.
4) $\gamma(U^s x) = 0$ for all $\gamma \in \mathrm{im}(D_A+N)'$ and $s = 0,\dots,k-1$.

From this proposition one concludes, in particular, the following fact we shall often use.

REMARK 6.1. $x \in \mathcal{A}_{k+1}$ if and only if $x, Ux \in \mathcal{A}_k$.

Now we introduce numbers α_k called α-numbers of A:

$$\alpha_k := \dim \mathcal{A}_k / \mathcal{A}_{k+1} \quad (= \dim \mathcal{A}_k - \dim \mathcal{A}_{k+1}).$$

Obviously,

$$\dim \mathcal{A}_k = \sum_{i=k}^{q} \alpha_i . \tag{6.1}$$

PROPOSITION 6.2. The sequence $\{\alpha_k\}$ is non-increasing.

PROOF. The subspace $\mathcal{A}_k$ admits a representation $\mathcal{A}_k = \mathcal{A}_{k+1} \oplus \mathrm{lin}\{x_1,\dots,x_\alpha\}$[1], where $\alpha = \alpha_k$ and the vectors x_i are linearly independent. The vectors Ux_i belong to $\mathcal{A}_{k-1}$ but not to $\mathcal{A}_k$. Let us assume now for a moment that $\alpha_{k-1} < \alpha$. Then there exists a vector $\lambda = (\lambda_i)_1^\alpha \neq 0$ such that $\sum_{i=1}^{\alpha} \lambda_i Ux_i \in \mathcal{A}_k$. Putting $x := \sum_{i=1}^{\alpha} \lambda_i x_i$ we obtain $x, Ux \in \mathcal{A}_k$, hence $x \in \mathcal{A}_{k+1}$, by Remark 6.1. This implies $x = 0$ and $\lambda = 0$, which is a contradiction to our assumption $\lambda \neq 0$. Therefore, $\alpha_{k-1} \geq \alpha$. ≡

PROPOSITION 6.3. Suppose $A \in \mathcal{T}_r^o$, then $\alpha_k \leq r$ $(k = 0,1,2,\dots)$.

PROOF. In view of Prop. 6.2, it remains to prove $\alpha_o \leq r$. Let D denote the restriction of $D_A + N$ onto $\ker A$. Then we have $\ker D = \mathcal{A}_1$, according to (3'). Hence,

$$\alpha_o = \dim \mathcal{A}_o / \mathcal{A}_1 = \mathrm{rank}\, D \leq \mathrm{rank}(D_A + N) = r,$$

and the proposition is proved. ≡

Now we deal with decompositions of $\ker A$ of the form

$$\ker A = \mathcal{C}_o \oplus \mathcal{C}_1 \oplus U\mathcal{C}_1 \oplus \dots \oplus \mathcal{C}_p \oplus U\mathcal{C}_p \oplus \dots \oplus U^p\mathcal{C}_p . \tag{6.2}$$

Obviously, having such a decomposition, bases consisting of (U,N)-chains can be easily constructed. The sequence of numbers γ_i, $\gamma_i := \sum_{j=1}^{p} \dim \mathcal{C}_i$ will be called characteristic of the decomposition (6.2). We are interested, of course, in decompositions (6.2) with minimal p.

PROPOSITION 6.4. There exists a decomposition (6.2) with characteristic $\{\alpha_i\}$ $(i = 0,\dots,q)$ and

$$\mathcal{C}_k \subseteq \mathcal{A}_k, \quad \mathcal{C}_k \cap \mathcal{A}_{k+1} = \{0\}. \tag{6.3}$$

PROOF. We put first $\mathcal{C}_q := \mathcal{A}_q$, then $\mathcal{C}_q + U\mathcal{C}_q \subseteq \mathcal{A}_{q-1}$. We show that $\mathcal{C}_q \cap U\mathcal{C}_q = \{0\}$. Suppose $x \in \mathcal{C}_q \cap U\mathcal{C}_q$, then $x = Uz$ for some $z \in \mathcal{C}_q$.

---

[1] Here and in all what follows ⊕ denotes the direct sum.

This implies $z \in \mathcal{A}_{q+1}$, in view of Remark 6.1, consequently $z = 0$ and $x = 0$. Thus, we obtained that the sum $\mathcal{C}_q + U\mathcal{C}_q$ is direct. Since, furthermore, $\dim U\mathcal{C}_q = \dim \mathcal{C}_q$, we have

$$\dim \mathcal{A}_{q-1} | \mathcal{C}_q \oplus U\mathcal{C}_q = \alpha_{q-1} - \alpha_q.$$

Hence, there is a direct complement $\mathcal{C}_{q-1}$ of $\mathcal{C}_q \oplus U\mathcal{C}_q$ in $\mathcal{A}_{q-1}$ with $\dim \mathcal{C}_{q-1} = \alpha_{q-1} - \alpha_q$.

Now, in the same manner subspaces $\mathcal{C}_{q-2}, \dots, \mathcal{C}_0$ satisfying (6.3) will be constructed. We explain the construction of $\mathcal{C}_{q-n-1}$ supposing $\mathcal{C}_q, \dots, \mathcal{C}_{q-n}$ are known and

$$\mathcal{A}_k = \mathcal{C}_k \oplus \mathcal{C}_{k+1} \oplus U\mathcal{C}_{k+1} \oplus \dots \oplus \mathcal{C}_q \oplus \dots \oplus U^{q-k}\mathcal{C}_q \qquad (6.2')$$

holds for $k = q-n$. We have $\mathcal{A}_{q-n} + U\mathcal{A}_{q-n} \subseteq \mathcal{A}_{q-n-1}$ and show $U^{n+1-i}\mathcal{C}_{q-i} \cap \mathcal{A}_0 = \{0\}$ $(i = 0,\dots,n)$. Suppose x belongs to this intersection. Then $x = U^{n+1-i}z$ for some $z \in \mathcal{C}_{q-i}$. This yields $z \in \mathcal{A}_{q-k+1}$, consequently, $z = 0$ and $x = 0$. We conclude that $\mathcal{A}_{q-n} + U\mathcal{A}_{q-n}$ is the direct sum of $\mathcal{A}_{q-n}$ and $U^{n+1-i}\mathcal{C}_{q-i}$ $(i = 0,\dots,n)$. By an elementary calculation using (6.1) the equality $\dim \mathcal{A}_{q-n-1} | \mathcal{A}_{q-n} + U\mathcal{A}_{q-n} = \alpha_{q-n-1} - \alpha_{q-n}$ is obtained. The subspace $\mathcal{C}_{q-n-1}$ is a direct complement of $\mathcal{A}_{q-n} + U\mathcal{A}_{q-n}$ in $\mathcal{A}_{q-n-1}$ now. Therefore, we have a representation (6.2') for $k = q-n-1$.

After q steps of this construction a representation (6.2) is obtained. $\equiv$

Decompositions (6.2) of $\ker A$ with characteristic $\{\alpha_i\}$ (i.e. such as constructed in the proof of Prop. 6.4) will be referred to as <u>canonical decompositions.</u> We are going now to compare canonical decompositions with other ones.

<u>PROPOSITION 6.5.</u> Suppose by (6.2) a decomposition with the characteristic $\{\gamma_i\}$ $(i = 1,\dots,p)$ is given. Then the following estimations hold

$$\alpha_0 + \dots + \alpha_{i-1} - j\alpha_i \le \gamma_0 + \dots + \gamma_{i-1} - j\gamma_i \qquad (6.4.)$$
$$(j = 0,\dots,i;\ i = 1,2,\dots).$$

<u>PROOF.</u> The following inclusions hold

$$\tilde{\mathcal{A}}_i := \mathcal{C}_i \oplus \mathcal{C}_{i+1} \oplus U\mathcal{C}_{i+1} \oplus \dots \oplus \mathcal{C}_p \oplus \dots \oplus U^{p-i}\mathcal{C}_p \subseteq \mathcal{A}_i.$$

This implies

$$\tilde{\mathcal{A}}_i + U\tilde{\mathcal{A}}_i + \dots + U^j\tilde{\mathcal{A}}_i \subseteq \mathcal{A}_i + U\mathcal{A}_i + \dots + U^j\mathcal{A}_i.$$

Let $\tilde{\mathcal{A}}_{ij}$ be the subspace on the left and $\mathcal{A}_{ij}$ on the right side of the latter inclusion. We compute the numbers $\gamma_{ij} := \dim \tilde{\mathcal{A}}_{ij}$.

Obviously, $\gamma_{io} = \gamma_i + \dots + \gamma_p$ and $\gamma_{i,j+1} = \gamma_{ij} + \gamma_i$, hence $\gamma_{ij} = \gamma_{io} + j\gamma_i$. This implies $\gamma_{ij} = \dim \ker A - \gamma_o - \dots - \gamma_{i-1} + j\gamma_i$. Analogously, we have $\dim \mathcal{A}_{ij} = \dim \ker A - \alpha_o - \dots - \alpha_{i-1} + j\alpha_i$. Comparing the latter two equalities we obtain the estimation (6.4). ≡

A consequence of Prop. 6.5 is

PROPOSITION 6.6. Suppose by (6.2) a decomposition with the characteristic $\{\gamma_i\}$ is given. Then $\gamma_o = \alpha_o$ implies $\{\gamma_i\} = \{\alpha_i\}$, i.e. the decomposition is canonical.

PROOF. We assume $\gamma_j = \alpha_j$ if $j = 0,\dots,i-1$ and prove $\gamma_i = \alpha_i$. According to (6.4) we have

$$\alpha_o + \dots + \alpha_i \le \gamma_o + \dots + \gamma_i,$$

which implies $\alpha_i \le \gamma_i$, and, on the other hand

$$\alpha_o + \dots + \alpha_{i-1} - \alpha_i \le \gamma_o + \dots + \gamma_{i-1} - \gamma_i,$$

which implies $-\alpha_i \le -\gamma_i$, hence $\alpha_i = \gamma_i$. ≡

PROOF OF THEOREM 6.1. Let (6.2) be a canonical decomposition of $\ker A$. In any subspace $\mathcal{C}_i$ we choose a basis $\{x_{ij} : j=1,\dots,\alpha_i - \alpha_{i+1}\}$. For fixed k and i the system $\{U^k x_{ij} : j=1,\dots,\alpha_i - \alpha_{i+1}\}$ is linearly independent, because $\sum_j \lambda_j U^k x_{ij} = 0$ implies $x := \sum_j \lambda_j x_{ij} \in \mathcal{A}_k$ for all k, hence $x = 0$. We conclude that $\{U^k x_{ij} : i=0,\dots,q;\ j=1,\dots,\alpha_i - \alpha_{i+1};\ k=0,\dots,i\}$ is a basis of $r_o = \alpha_o$ (U,N)-chains.

We assume now that $\{U^k \tilde{x}_{ij} : i=0,\dots,p;\ j=1,\dots,\gamma_i - \gamma_{i+1};\ k=0,\dots,i\}$ is another basis of α_o (U,N)-chains. Then putting $\tilde{\mathcal{C}}_i := \mathrm{lin}\{x_{ij}\}$ we obtain a canonical decomposition of $\ker A$. Applying Prop. 6.6 we conclude $p = q$ and $\gamma_i = \alpha_i$. This proves the theorem. ≡

COROLLARY 6.1. The minimal number of (U,N)-chains spanning $\ker A$ equals α_o.

A basis of (U,N)-chains corresponding to a canonical decomposition will be called canonical basis.

Analogously to Theorem 6.1 we have

THEOREM 6.1'. Suppose $A \in \mathcal{T}_r^o$, then there is a basis of $\ker A'$ consisting of $\beta_o \le r$ (V',N')-chains. For the minimal value of β_o the lengths of the chains are uniquely determined by A.

The lengths of the (V',N')-chains in a canonical basis of $\ker A'$ are determined by the subspaces $\mathcal{B}_k := \ker[\mathrm{Con}_k A\ \mathrm{Con}_{k-1} N]'$ and the numbers $\beta_k := \dim \mathcal{B}_k | \mathcal{B}_{k+1}$. These numbers will be called β-numbers of A. Clearly, the β-numbers are the α-numbers of A'.

We mention an analogue of Prop. 6.1, which is a consequence of the normal solvability of A'.

PROPOSITION 6.1'. The functional $\xi \in \ker A'$ belongs to $\mathcal{B}_k$ if and only if $\xi(V^s g) = 0$ for all $g \in \mathrm{im}(D_A + N)$ and $s = 0,\dots,k-1$.

6.2. Duality. We know from I, Section 5 that the numbers α_0 and β_0 of U-chains spanning ker H and ker H^T, where H is Hankel, are not independent, but $\alpha_0 + \beta_0 \leq 2$. A similar statement is known in the theory of Wiener-Hopf operators. Now, we are going to generalize these particular results. Another problem we deal with is the question about the structure of ker A and ker A' on one side and the solvability properties of the fundamental equations on the other side. We begin with a first result concerning the latter problem.

We denote

$$\mathcal{G}_A := \mathrm{im}(D_A + N), \quad \mathcal{F}_A := \mathrm{im}(D_A + N)'.$$

Fundamental equations are, by definition, equations of the form

$$Ax = g \in \mathcal{G}_A, \qquad A'\xi = \gamma \in \mathcal{F}_A. \tag{6.5}$$

THEOREM 6.2. Suppose $A \in \mathcal{T}_r^o$, then

$$r = \beta_0 + \dim \mathcal{G}_A \cap \mathrm{im}\, A = \alpha_0 + \dim \mathcal{F}_A \cap \mathrm{im}\, A'.$$

In other words, the number of (V',N')-chains spanning ker A' ((U,N)-chains spanning ker A) plus the number of linearly independent solvable fundamental equations $Ax = g$ ($A'\xi = \gamma$) equals r.

PROOF. We have, by definition, a representation $\ker A' = \mathcal{B}_1 \oplus \mathcal{H}$, where $\dim \mathcal{H} = \beta_0$. Let $\{\varphi_i\}$ be a basis of $\mathcal{H}$ and $\Phi : \mathcal{G}_A \to \mathbb{C}^{\beta_0}$ the operator $\Phi g := (\varphi_i(g))_1^{\beta_0}$. We show that Φ is an epimorphism. Suppose $\lambda = (\lambda_i)_1^{\beta_0}$ is orthogonal to im Φ. Then, putting $\varphi := \sum \lambda_i \varphi_i$, we have $\varphi(g) = 0$ for all $g \in \mathcal{G}_A$. According to Prop. 6.1' this implies $\varphi \in \mathcal{B}_1$, hence $\varphi = 0$. Consequently, $\lambda = 0$, i.e. Φ is an epimorphism.

Now we prove the equality

$$\ker \Phi = \mathcal{G}_A \cap \mathrm{im}\, A. \tag{6.6}$$

As noted above the operators from $\mathcal{T}_r^o$ are Fredholm. Therefore, by Prop. 0.1, g belongs to im A if and only if $\varphi(g) = 0$ for all $\varphi \in \ker A'$. In virtue of Prop. 6.1, $\varphi(g) = 0$ holds for all $\varphi \in \mathcal{B}_1$ and $g \in \mathcal{G}_A$. Hence, $g \in \mathcal{G}_A$ belongs to im A if and only if $\Phi g = 0$. This proves (6.6). We obtain

$$r = \dim \mathcal{G}_A = \dim \ker \Phi + \dim \mathrm{im}\, \Phi = \dim \mathcal{G}_A \cap \mathrm{im}\, A + \beta_0,$$

and the theorem is proved. ≡

Let us note that Theorem 6.2 is a generalization of the invertibility criteria proved in Section 1 (Theorem 1.1).

THEOREM 6.3. Suppose $A \in \mathcal{T}_r^o$, then

$$\alpha_0 + \beta_0 \leq r. \tag{6.7}$$

PROOF. Let D be the restriction of $D_A + N$ onto $\ker A$. Then, by definition, $\ker D = \mathcal{A}_1$ and $\operatorname{im} D \subseteq \mathcal{G}_A \cap \operatorname{im} A$. Therefore,

$$\alpha_0 = \dim \ker A / \mathcal{A}_1 = \dim \operatorname{im} D \leq \dim \mathcal{G}_A \cap \operatorname{im} A.$$

It remains to apply Theorem 6.2. $\equiv$

COROLLARY 6.2. Suppose $A \in \mathcal{T}_r^0$, then $\operatorname{ind} A < r$ implies $\alpha_0 \leq r-1$ and $\operatorname{ind} A > -r$ implies $\beta_0 \leq r-1$.

PROOF. If $\dim \ker A < r$, then the assertion is trivial. If $\dim \ker A \geq r$ then $\dim \ker A > \operatorname{ind} A$, hence $\dim \ker A' > 0$, consequently $\beta_0 > 0$. In view of (6.7) this yields $\alpha_0 \leq r-1$. Analogously the second inequality is proved. $\equiv$

6.3. Partial indices. The concept of partial indices plays an important role in the theory of matrix-valued Toeplitz and singular integral operators. In I, Section 5 we saw that this concept can be introduced for finite T- and H-matrices, too. In this subsection we give a definition of partial indices of operators $A \in \mathcal{T}_r^0$, which covers the particular definitions.

Let $\{x_l, Ux_l, \dots, U^{\nu_l - 1} x_l\}$ $(l = 1, \dots, \alpha_0)$ be a canonical basis of $\ker A$ and $\{\xi_l, V'\xi_l, \dots, (V')^{\mu_l - 1} \xi_l\}$ $(l = 1, \dots, \beta_0)$ a canonical basis of $\ker A'$. The $\alpha_0 + \beta_0 (\leq r)$ numbers ν_l, μ_l are, due to Theorem 6.1 and 6.1', uniquely determined by A.

We arrange the chains such that $\nu_1 \geq \dots \geq \nu_{\alpha_0}$ and $\mu_1 \geq \dots \geq \mu_{\beta_0}$.

DEFINITION 6.3. The system of partial indices of the operator $A \in \mathcal{T}_r^0$ is, by definition, the r-tuple of integers $\varkappa(A) := (\varkappa_i(A))_1^r$ defined by

$$\varkappa_i = \begin{cases} \nu_i & : i = 1, \dots, \alpha_0 \\ 0 & : i = \alpha_0 + 1, \dots, r - \beta_0 \\ -\mu_{r+1-i} & : i = r - \beta_0 + 1, \dots, r. \end{cases}$$

Clearly, $A \in \mathcal{T}_r^0$ is invertible if and only if $\varkappa(A) = 0$. Let us explain the connection between α- and β-numbers and the partial indices. The following relationship immediately follows from the definition.

REMARK 6.2.

$$\alpha_{i-1} - \alpha_i = |\{l : \varkappa_l = i\}| \text{ if } i > 0, \tag{6.8a}$$

$$\beta_{i-1} - \beta_i = |\{l : \varkappa_l = -i\}| \text{ if } i < 0, \tag{6.8b}$$

$$r - \alpha_0 - \beta_0 = |\{l : \varkappa_l = 0\}|. \tag{6.8c}$$

Here $|M|$ denotes the number of elements in M.

For the following it is convenient to introduce the designation

$$\alpha_{-i-1} := r - \beta_i, \quad \beta_{-i-1} := r - \alpha_i \quad (i = 0, 1, \dots). \tag{6.9}$$

Speaking in the following about α-numbers we have the numbers α_i for $i = 0, \pm 1, \ldots$ in mind.

PROPOSITION 6.7. For any integer k the following relations hold

$$\sum_{\varkappa_l > k} (\varkappa_l - k) = \sum_{i \geq k} \alpha_i, \quad \sum_{\varkappa_l < -k} (\varkappa_l + k) = - \sum_{i \geq k} \beta_i. \tag{6.10}$$

PROOF. Let $\{x_l, \ldots, U^{\varkappa_l - 1} x_l\}$ $(l = 1, \ldots, \alpha_o)$ be a canonical basis of $\ker A$. Then $(x_l, \ldots, U^{\varkappa_l - 1 - k} x_l\}$ $(l = 1, \ldots, \alpha_k)$ is a basis of $\mathcal{A}_k$ $(k = 1, 2, \ldots)$. This yields the first relation of (6.10) for $k \geq 0$. Analogously the second relation is obtained. Now we consider the case $k < 0$. Using (6.10) for $k \geq 0$ we conclude

$$\begin{aligned}\sum_{\varkappa_l > k} (\varkappa_l - k) &= \dim \ker A + \sum_{o > \varkappa_l > k} (\varkappa_l - k) \\ &= \dim \ker A - \sum_{i < -k} \beta_i - kr \\ &= \dim \ker A - \sum_{i=o}^{-k-1} (r - \beta_i).\end{aligned}$$

On the other hand, there is

$$\sum_{i \geq k} \alpha_i = \dim \ker A + \sum_{i=k}^{-1} \alpha_i.$$

Taking into account (6.9) we obtain the first relation of (6.10). The second relation is proved analogously. ≡

The following relations are particular cases of (6.10)

$$\operatorname{ind} A = \sum_{l=1}^{r} \varkappa_l, \quad \dim \ker A = \sum_{\varkappa_l > o} \varkappa_l, \quad \dim \ker A' = - \sum_{\varkappa_l < o} \varkappa_l. \tag{6.11}$$

We proceed the investigation of partial indices in Subsection 6.5.

6.4. Structure of the reduction.

In this subsection we analyse more in detail the structure of the finite-dimensional operator $D_A + N$. The corresponding results will be, in particular, generalizations of Theorem 6.2. We introduce subspaces $\mathcal{G}_k$ of $\mathcal{G}_A = \operatorname{im}(D_A + N)$ defining

$$\mathcal{G}_k := \{g \in \mathcal{G}_A : V^k g \in \operatorname{im}[\mathrm{Con}_k A \ \ \mathrm{Con}_{k-1} N]\}.$$

In particular, we have $\mathcal{G}_o = \mathcal{G}_A \cap \operatorname{im} A$, hence, by Theorem 6.2, $\dim \mathcal{G}_o = r - \beta_o$ of $A \in \mathcal{T}_r^o$. A generalization of this fact is

THEOREM 6.4. Suppose $A \in \mathcal{T}_r^o$, then

$$\dim \mathcal{G}_k = r - \beta_k \ \ (= \alpha_{-k-1}) \ \ (k = 0, 1, \ldots). \tag{6.12}$$

Before proving the theorem we note some preparatory facts. In virtue of the normal solvability of $[\mathrm{Con}_k A \ \ \mathrm{Con}_{k-1} N]$, a vector $g \in \mathcal{G}_A$

belongs to $\mathcal{G}_k$ if and only if $\gamma(V^k g) = 0$ for all $\gamma \in \mathcal{B}_k$. The subspace $\mathcal{B}_k$ can be represented as direct sum $\mathcal{B}_k = \mathcal{B}_{k+1} \oplus \mathcal{H}_k$ with $\mathcal{H}_k = \text{lin}\{\varphi_{ki}: i = 1,\dots,\beta_k\}$. By Prop. 6.1', we have $\gamma(V^k g) = 0$ for all $\gamma \in \mathcal{B}_{k+1}$. Therefore, g belongs to $\mathcal{G}_k$ if and only if $\varphi_{ki}(V^k g) = 0$ $(i = 1,\dots,\beta_k)$. In that way we proved

<u>LEMMA 6.1.</u> Let $\Phi_k V^k : \mathcal{G}_A \to \mathbb{C}^{\beta_k}$ be the operator defined by $\Phi_k V^k_g := ((\varphi_{ki}(g))_1^{\beta_k}$, then $\mathcal{G}_k = \ker \Phi_k V^k$.

<u>PROOF OF THEOREM 6.4.</u> We show that the operators $\Phi_k V^k$ are epimorphisms. Suppose $\lambda = (\lambda_i)_1^\beta$ $(\beta := \beta_k)$ is orthogonal to $\text{im}\,\Phi_k V^k$ and $\varphi := \sum_{i=1}^{\beta} \lambda_i \varphi_{ki} \in \mathcal{H}_k$. Then, by assumption, $\varphi(V^k g) = 0$ for all $g \in \mathcal{G}_A$, hence, by Prop. 6.1', $\varphi \in \mathcal{B}_{k+1}$. This implies $\varphi = 0$ and $\lambda = 0$. Taking into account Lemma 6.1 we obtain

$$r = \dim \mathcal{G}_A = \dim \ker \Phi_k V^k + \dim \text{im}\, \Phi_k V^k = \dim \mathcal{G}_k + \beta_k ,$$

and the theorem is proved. ≡

We introduce now subspaces $\mathcal{G}_{-1}, \mathcal{G}_{-2}, \dots$ defining

$$\mathcal{G}_{-k-1} := AU^{k+1}\mathcal{A}_k.$$

Obviously, we have also $\mathcal{G}_{-k-1} = (D_A + N)U^k \mathcal{A}_i$. Taking into account that, by definition, $\mathcal{A}_{k+1} = \mathcal{A}_k \cap \ker AU^{k+1}$ we obtain

$$\dim \mathcal{G}_{-k-1} = \alpha_k .$$

In virtue of (6.9) this yields the following result.

<u>REMARK 6.3.</u> The relation (6.12) holds for all integers k.

Analogously to the subspaces $\mathcal{G}_k$ the subspaces $\mathcal{F}_k$ of $\mathcal{F}_A = \text{im}(D_A+N)'$ will be introduced. We define

$$\mathcal{F}_k = \begin{cases} \{\gamma \in \mathcal{F}_A : (U')^k \gamma \in \text{im}[\text{Con}_k A \quad \text{Con}_{k-1} N]\} & : k \geq 0 \\ A'(V')^{-k}\mathcal{B}_{-k-1} & : k < 0. \end{cases}$$

Then, analogously to (6.12), the relation

$$\dim \mathcal{F}_k = r - \alpha_k = \beta_{-k-1} \tag{6.13}$$

holds for all integers k. We show now that the subspaces $\mathcal{F}_k$ and $\mathcal{G}_{-k-1}$ are closely connected. For this we make use of the following notation:

$$\mathcal{F}^\perp := \{x \in \mathbb{E} : \gamma(x) = 0 \text{ for all } \gamma \in \mathcal{F}\} \quad (\mathcal{F} \subseteq \mathbb{E}')$$

$$\mathcal{G}^\perp := \{\eta \in \mathbb{F}' : \eta(g) = 0 \text{ for all } g \in \mathcal{G}\} \quad (\mathcal{G} \subseteq \mathbb{F}).$$

PROPOSITION 6.8. Suppose $A \in \mathcal{T}_r^0$, then for all integers k

$$(D_A + N)\mathcal{F}_k^{\perp} = \mathcal{G}_{-k-1}, \quad (D_A + N)' \mathcal{G}_k = \mathcal{F}_{-k-1}. \tag{6.14}$$

PROOF. First we prove (6.14) for $k \geq 0$. In view of the normal solvability of $[\mathrm{Con}_k A \ \ \mathrm{Con}_{k-1} N]$, we have

$$\mathcal{F}_k = \{\gamma \in \mathcal{F}_A : \gamma(U^k x) = 0 \ \text{ for all } \ x \in \mathcal{A}_k\}. \tag{6.15}$$

Hence $\mathcal{F}_k = \mathcal{F}_A \cap (U^k \mathcal{A}_k)^{\perp}$, which implies $\mathcal{F}_k^{\perp} = \mathcal{F}_A^{\perp} + U^k \mathcal{A}_k =$ $= \ker(D_A + N) + U^k \mathcal{A}_k$. This leads to the first relation of (6.14). The second relation is proved analogously.

Now we consider the case $k < 0$. We have in this case $\mathcal{F}_k = (V')^{-k-1}(D_A + N)' \mathcal{B}_{-k-1}$. Hence $x \in \mathcal{F}_k^{\perp}$ if and only if $\gamma(V^{-k-1}(D_A + N)x) = 0$ for all $\gamma \in \mathcal{B}_{-k-1}$. The latter condition is equivalent to the fact $(D_A + N)x$ is orthogonal to $(V')^{-k-1}\mathcal{B}_{-k-1}$. Taking into account the normal solvability of $[\mathrm{Con}_{-k-1} A \ \ \mathrm{Con}_{-k} N]$ and the definition of $\mathcal{G}_{-k-1}$ we conclude the first relation of (6.14). The second relation is proved analogously. ≡

We define now the concept of a canonical basis of $\mathcal{G}_A$. Suppose $\mathcal{H}_k$ is a direct complement of $\mathcal{G}_{-k-1}$ in $\mathcal{G}_{-k}$ and g_{ki} $(i = 1,\dots,\alpha_{k-1} - \alpha_k)$ a basis of $\mathcal{H}_k$. Then the system of all g_{ki} is a basis of $\mathcal{G}_A$ called <u>canonical basis of</u> $\mathcal{G}_A$. Analogously, a canonical basis of $\mathcal{F}_A$ is, by definition, a basis $\{\gamma_{ki}\}$ such that $\mathrm{lin}\{\gamma_{ki} : i = 1,\dots,\beta_{k-1} - \beta_k\}$ is a direct complement of $\mathcal{F}_{-k-1}$ in $\mathcal{F}_{-k}$.

REMARK 6.4. Let $\{U^j x_{ki}\}$ be a canonical basis (consisting of (U,N)-chains) of $\ker A$. Then g_{ki} $(k > 0)$ can be chosen as $g_{ki} = (D_A + N)U^k x_{ki}$. Analogously, γ_{ki} $(k > 0)$ can be chosen as $\gamma_{ki} = (D_A + N)'(V')^k \xi_{ki}$, where $\{(V')^j \xi_{ki}\}$ is a canonical basis of $\ker A'$.

From Prop. 6.8 we conclude the following result.

THEOREM 6.5. Suppose $A \in \mathcal{T}_r^0$, then $D_A + N$ admits a representation

$$D_A + N = \sum_{k,i} \gamma_{-k,i}(\cdot) g_{ki}, \tag{6.16}$$

where $\{g_{ki}\}$, $\{\gamma_{ki}\}$ are canonical bases of $\mathcal{G}_A$, $\mathcal{F}_A$, respectively.

In (6.16) the double index (k,i) can be replaced by the index $\varkappa_l$, where $\varkappa_l$ runs over the system of partial indices. That means,

$$D_A + N = \sum_{l=1}^{r} \gamma_{-\varkappa_l}(\cdot) g_{\varkappa_l}, \tag{6.16'}$$

where

$$g_{\varkappa_1} \in \mathcal{G}_{-\varkappa_1} - \mathcal{G}_{-\varkappa_1 - 1}, \quad \gamma_{-\varkappa_1} \in \mathcal{F}_{\varkappa_1} - \mathcal{F}_{\varkappa_1 - 1} .$$

COROLLARY 6.3. Suppose $A \in \mathcal{T}_r^o$ and let α be an integer with $\alpha_0 \le \alpha \le \alpha_{-1}$. Then there are $x_i \in \mathbb{E}$, $\gamma^i \in \mathcal{F}_A$ $(i = 1,\dots,\alpha)$ and $\xi_j \in \mathbb{F}'$, $g^j \in \mathcal{G}_A$ $(j = 1,\dots,r-\alpha)$

$$(V - \sum_{j=1}^{r-\alpha} \xi_j(\cdot)g^j)A = A(U - \sum_{i=1}^{\alpha} \gamma^i(\cdot)x_i) + N. \tag{6.17}$$

The relation (6.17) follows from the fact that in (6.16) we have $g_{ki} \in \mathcal{G}_0 \subseteq \text{im } A$ if $k \ge 0$ and $\gamma_{ki} \in \mathcal{F}_0 \subseteq \text{im } A'$ if $k \ge 0$.

We show now that, in generalization to the corresponding fact for Hankel matrices (c.f. I, Prop. 5.9), the relation $g \in \mathcal{G}_k$ is equivalent to the solvability of some generalized fundamental equation. Let $\mathcal{G}^{(k)}$ denote the subspace $\mathcal{G}_A + V\mathcal{G}_A + \dots + V^{k-1}\mathcal{G}_A$ $(\mathcal{G}^{(-1)} := \{0\})$. <u>Fundamental equations of order</u> $k \ge 0$ are, by definition, equations of the form

$$Ax = V^k g + g_k, \tag{6.18a}$$

where $g \in \mathcal{G}_A$ is fixed and $g_k \in \mathcal{G}^{(k)}$ is arbitrary. <u>Fundamental equations of order</u> <u>$k < 0$</u> are, by definition equations of the form

$$AU^{-k}x = g, \quad x \in \mathcal{A}_{-k-1}. \tag{6.18b}$$

<u>THEOREM 6.6.</u> The vector $g \in \mathcal{G}_A$ belongs to $\mathcal{G}_\varkappa$ if and only if (6.18) is solvable for $k = \varkappa$.

<u>PROOF.</u> In case $\varkappa \le 0$ the assertion immediately follows from the definition of the subspaces $\mathcal{G}_k$. Suppose now $\varkappa > 0$. Let the operators Φ_k be defined as above (c.f. Lemma 6.1). Assuming (6.18a) is solvable if $k = \varkappa$ we obtain $\Phi_s(V^\varkappa g + g_\varkappa) = 0$ $(s = 0,1,\dots)$. Since $\Phi_\varkappa g_\varkappa = 0$, we conclude, taking into account Lemma 6.1, $g \in \mathcal{G}_\varkappa$.
Now suppose $g \in \mathcal{G}_\varkappa$. We define vectors $g_0,\dots,g_{\varkappa-1}$ by

$$\begin{array}{lcl}
\Phi_{\varkappa-1} V^{\varkappa-1} g_{\varkappa-1} & = & -\Phi_{\varkappa-1} V^\varkappa g \\
\Phi_{\varkappa-2} V^{\varkappa-1} g_{\varkappa-1} + \Phi_{\varkappa-2} V^{\varkappa-2} g_{\varkappa-2} & = & -\Phi_{\varkappa-2} V^\varkappa g \\
\vdots \qquad\qquad \ddots & & \vdots \\
\Phi_0 V^{\varkappa-1} g_{\varkappa-1} + \dots + \Phi_0 g_0 & = & -\Phi_0 V^\varkappa g .
\end{array}$$

The latter system is solvable, since $\Phi_k V^k$ are epimorphisms. Putting $g_\varkappa = \sum_{i=0}^{\varkappa-1} V^i g_i$ we obtain $\Phi_s(V^\varkappa g + g_\varkappa) = 0$ $(s = 0,1,\dots)$. Hence $V^\varkappa g + g_\varkappa \in \text{im } A$, which means (6.18a) is solvable. ≡

We remark, finally in this subsection, that the numbers β_k $(k=0,1,\dots)$ (characterizing the structure of $\ker A' \subseteq \mathbb{F}$) can also expressed as the dimension of some subspaces of $\mathbb{B}$. We define

$$\mathcal{A}_{-k-1} := \{x \in \mathbb{B} : Ax \in \mathcal{G}^{(k)}\}.$$

PROPOSITION 6.9. Suppose $A \in \mathcal{T}_r^o$, then

$$\dim \mathcal{A}_{-k-1}|\mathcal{A}_{-k} = r - \beta_k - \dim \mathcal{G}^{(k)} \ (= \alpha_{-k-1} - \dim \mathcal{G}^{(k)}).$$

6.5. Half-stability of partial indices. Let Z_r denote the class of all non-increasing sequences $(n_i)_1^r$ of integers. We define the half-ordering $(n_i)_1^r \geq (m_i)_1^r$ by

$$\sum_{i=1}^{k} n_i \geq \sum_{i=1}^{k} m_i \quad (k = 1,\dots,r-1), \quad \sum_{i=1}^{r} n_i = \sum_{i=1}^{r} m_i.$$

We show that the system of partial indices $\varkappa(A)$ of a TLO A is not stable, in general, relative to small normperturbations but half-stable with respect to " $\leq$ ". More precisely speaking we have

THEOREM 6.7. Suppose $A \in \mathcal{T}_r^o$, then there exists a $\varepsilon > 0$ such that for all operators $\tilde{A} \in \mathcal{T}_r^o$ with $\|A - \tilde{A}\| < \varepsilon$ the estimation

$$\varkappa(\tilde{A}) \leq \varkappa(A)$$

holds.

The converse of this result is

THEOREM 6.8. Suppose $A \in \mathcal{T}_r^o$ and $\tilde{\varkappa} \in Z_r$ is such that $\tilde{\varkappa} \leq \varkappa(A)$. Then for all $\varepsilon > 0$ there exists an operator $\tilde{A} \in \mathcal{T}_r^o$ such that $\tilde{A} - A$ is finite-dimensional, $\|\tilde{A} - A\| < \varepsilon$ and $\varkappa(\tilde{A}) = \tilde{\varkappa}$.

COROLLARY 6.4. The system of partial indices $\varkappa(A) = (\varkappa_i)_1^r$ of $A \in \mathcal{T}_r^o$ is stable with respect to small norm perturbations if and only if $|\varkappa_i - \varkappa_j| \leq 1$ for all i and j.

Now we are going to prove the theorems. Suppose $\varkappa = (\varkappa_i)_1^r$, $\tilde{\varkappa} = (\tilde{\varkappa}_i)_1^r \in Z_r$. We shall say that $\tilde{\varkappa}$ is obtained from $\varkappa$ by an elementary transformation if for some indices p, q the following relations hold

$$\varkappa_p > \varkappa_{p+1} > \varkappa_{q-1} > \varkappa_q, \tag{6.19}$$
$$\tilde{\varkappa}_i = \varkappa_i \quad (i \neq p,q), \quad \tilde{\varkappa}_p = \varkappa_p - 1, \quad \tilde{\varkappa}_q = \varkappa_q + 1.$$

It is easily seen that the inequality $\tilde{\varkappa} \leq \varkappa$ holds if and only if $\tilde{\varkappa}$ is obtained from $\varkappa$ by a finite number of elementary transformations. This fact can be used to prove the following elementary property.

LEMMA 6.2. Let $\varkappa = (\varkappa_i)_1^r$, $\tilde{\varkappa} = (\varkappa_i)_1^r$ belong to Z_r and

$$a_k := \sum_{\varkappa_i > k} (\varkappa_i - k), \qquad \tilde{a}_k := \sum_{\tilde{\varkappa}_i > k} (\tilde{\varkappa}_i - k). \tag{6.20}$$

Then the inequality $\tilde{\varkappa} \leq \varkappa$ holds if and only if $\tilde{a}_k \leq a_k$ for all $k \in \mathbb{Z}$.

PROOF OF THEOREM 6.7. Let $\varkappa = \varkappa(A)$, $\tilde{\varkappa} = \varkappa(\tilde{A})$ be the systems of partial indices of the operators A, $\tilde{A}$, respectively. Then the numbers a_k defined by (6.20) are the dimensions of the subspaces $\mathcal{A}_k$ if $k = 0,1,\dots$. In the case of $k < 0$ we have, in view of Prop. 6.7, $a_k = \operatorname{ind} A - kr + \dim B_{-k}$. Analogous relations hold for $\tilde{a}_k$ $(k \in \mathbb{Z})$. By assumption, the operators $\begin{bmatrix} \mathrm{Obs}_k A \\ \mathrm{Obs}_{k-1} N \end{bmatrix}$ are Φ_+- and $[\mathrm{Con}_k A \;\; \mathrm{Con}_{k-1} N]$ are Φ_--operators. Furthermore, a small norm pertubation of A yields a small norm pertubation of the corresponding observability and controllability operators. Therefore, Prop. 0.3 can be applied. We obtain that there exists an $\varepsilon > 0$ such that $\|\tilde{A} - A\| < \varepsilon$ implies $\tilde{a}_k \leq a_k$. In virtue of Lemma 6.2, this yields $\tilde{\varkappa} \leq \varkappa$. ≡

REMARK 6.5. The proof of Theorem 6.7 shows that it is not necessary to assume $\tilde{A} \in \mathcal{T}_r^o$ in the assertion of the theorem, if one defines the concept of partial indices for operators $\tilde{A}$, for which $\alpha_o + \beta_o \leq r$.

We discuss now the PROOF OF THEOREM 6.8. Because it is rather technical, we restrict ourselves to present the idea of the proof.
Obviously, it is sufficient to prove the theorem for the case when $\tilde{\varkappa}$ is obtained from $\varkappa(A)$ by an elementary transformation. Let us assume that (6.19) holds.
The main instrument for proving Theorem 6.8 is the canonical decomposition of $D_A + N$ (Theorem 6.5). Suppose this operator admits the representation (6.16'). We put

$$\tilde{A} := A - \varepsilon\, \gamma_{-\varkappa_p}(\cdot) g_{\varkappa_q} .$$

Then we have

$$(D_{\tilde{A}} + N)x = \sum_{l \neq p,q} \gamma_{-\varkappa_l}(x) g_{\varkappa_l} + (\gamma_{-\varkappa_q} - \varepsilon U' \gamma_{-\varkappa_p})(x) g_{\varkappa_q}$$
$$+ \gamma_{\varkappa_p}(x)(g_{\varkappa_p} + \varepsilon V g_q).$$

Therefore, $\tilde{A}$ belongs to $\mathcal{T}_r$ for sufficiently small ε. Now, it can be shown that $\tilde{\varkappa}$ is just the system of partial indices of the operator $\tilde{A}$ if ε is sufficiently small (one has to distinguish the three cases: $\varkappa_p > 0 > \varkappa_q$, $\varkappa_p > \varkappa_q = 0$ and $\varkappa_p > \varkappa_q > 0$). This proves the theorem. ≡

6.6. Application to the factorization problem. Recently the approach presented in this section has been applied for solving the problem of factorization of matrix functions in algebras of bounded functions (see G. HEINIG, B. SILBERMANN [1]). In this subsection we formulate the factorization problem and the main result. We renounce to present the results as well as their proofs more in detail, since they are contained in the monograph A. BÖTTCHER, B. SILBERMANN [1] published in the same series "Mathematical Research" recently.

Let $\mathcal{A}$ be a Banach algebra of complex valued measurable essential bounded functions on $\mathbb{T}$ [1] containing the functions $\varphi(t) = t^k$ $(k = 0, \pm 1)$.
The algebra is said to be <u>decomposing</u> if $P\mathcal{A}$ and $Q\mathcal{A}$ are subalgebras of $\mathcal{A}$. If $\mathcal{A}$ is decomposing, then we set

$$\mathcal{A}_+ := P\mathcal{A}, \quad \overset{\circ}{\mathcal{A}}_- := Q\mathcal{A}, \quad \mathcal{A}_- := \mathcal{A}_0 + \overset{\circ}{\mathcal{A}}_-,$$

where $\mathcal{A}_0$ the class of constant functions. The group of invertible elements of an algebra $\mathcal{B}$ will be denoted by $G\mathcal{B}$.

<u>DEFINITION 6.4.</u> A matrix function $a(t) \in \mathcal{A}^{n\times n}$ is said to admit a (right) <u>factorization in $\mathcal{A}$</u> if there is a representation

$$a(t) = a_-(t)d(t)a_+(t) \quad (t \in \mathbb{T}),$$

where $a_\pm(t) \in G\mathcal{A}_\pm^{n\times n}$ and $d(t) = \operatorname{diag}(t^{\nu_1},\dots,t^{\nu_n})$ $(\nu_i \in \mathbb{Z},\ \nu_1 \le \dots \le \nu_n)$ we shall say that $\mathcal{A}$ has the <u>n-factorization property</u> if any matrix function $a(t) \in G\mathcal{A}^{n\times n}$ admits a factorization in $\mathcal{A}$.
The factorization problem we deal with is the problem whether a given algebra has the n-factorization property.
With any matrix function $a(t) \in \mathcal{A}^{n\times n}$ we associate two Toeplitz operators

$$T_a^+ \varphi_+ = Pa(t)\, \varphi_+(t) \qquad (\varphi_+ \in \mathcal{A}_+^n)$$

and

$$T_a^- \varphi_- = Qa(t)^T \varphi_-(t) \quad (\varphi_- \in \overset{\circ}{\mathcal{A}}{}_-^n).$$

It is easily verified that $\overset{\circ}{\mathcal{A}}{}_-^n$ is invariant under $(T_a^+)'$ and the restriction of $(T_a^+)'$ onto $\overset{\circ}{\mathcal{A}}{}_-^n$ coincides with T_a^-, provided that the dual operator is defined via the bilinear form

$$(\varphi_+,\varphi_-) = \int \varphi_-(t)^T \varphi_+(t)dt \quad (\varphi_\pm \in \mathcal{A}_\pm^n).$$

Suppose $a(t)$ admits a factorization in $\mathcal{A}$. Then, obviously, T_a^+ and T_a^- are Fredholm. Moreover, it is easily seen that the system of partial indices of T_a^+ (in the sense of 6.3) is just $(-\nu_1,\dots,-\nu_n)$ and of T_a^- is $(\nu_n,\dots,\nu_1)$.
The main part of the paper G. HEINIG, B. SILBERMANN [1] is concerned with the proof that the converse is also true. This leads to the following result.

<u>THEOREM 6.9.</u> The algebra $\mathcal{A}$ has the n-factorization property for all $n = 1,2,\dots$ if and only if for all $a(t) \in G\mathcal{A}^{n\times n}$ the following conditions are satisfied:

a) T_a^+ and T_a^- are Fredholm

b) $\operatorname{ind} T_a^+ + \operatorname{ind} T_a^- = 0$.

From Theorem 6.9 it can be deduced simple sufficient conditions for the

---

[1] The unit circle can be replaced by a closed Liapunov contour.

n-factorization property. A simple non-classical example of an algebra with n-factorization property is the algebra $\Lambda_\alpha + H_\infty$, where Λ_α denotes the algebra of Hölder continuous functions $(0 < \alpha < 1)$.

<u>PROBLEM.</u> Does the 1-factorization property imply the n-factorization property for all n?

<u>6.7. Examples.</u> In this subsection we give a few further examples for the application of the abstract results proved in this section. Let us note that we present only some immediate corollaries of them. A more detailed analysis of the special classes of TLO will certainly yield stronger results. In that way we hope to generalize some further statements of the theory of Toeplitz- and Hankel matrices and the Wiener-Hopf theory as well, for example the inversion algorithms presented in I, Section 6, to the block case and to matrices close to Toeplitz.

Our first remark is that, obviously, the kernel structure properties of Toeplitz matrices proved in I, Section 6 are corollaries of the general results indeed. Three generalizations of this fact are the following statements.

<u>COROLLARY 6.5.</u> The kernel of an mxn block Toeplitz matrix with entries $a_{i-k} \in \mathbb{C}^{d\times d}$ and their transposed is spanned by $\alpha_0 \leq 2d$ (U_n, P_{n-1})-chains, respectively $\beta_0 \leq 2d$ (U_m, P_{m-1})-chains, where $\alpha_0 + \beta_0 \leq 2d$. Here U_n denotes the block-wise forward shift in $\mathbb{C}^{nd}$ and P_l is the projection onto the l th block component.

<u>COROLLARY 6.6.</u> The kernel of an nxn paired T-matrix (2.1) is spanned by $\alpha_0 \leq 2$ $(U_l + U_{n-l}, P_{l-1,n-1})$-chains, the kernel of the transposed by $\beta_0 \leq 2$ (U_n, P_{n-1})-chains where $\alpha_0 + \beta_0 \leq 3$. Here $P_{l-1,n-1}$ denotes the projection onto the (l-1)th and (n-1)th components.

<u>COROLLARY 6.7.</u> Suppose $A = T + H \in \mathbb{C}^{n\times n}$, where T is Toeplitz and H is Hankel. Then ker A + ker(T - H) is spanned by $\alpha_0 \leq 2$ (U_n, P_{n-1})-chains.

Analogously, kernel structure properties for matrices closed to Vandermonde and matrices with small rank diagonal reduction can be formulated. A particular consequence is the fact that square matrices of the form (2.44) are always regular. Another result is that the kernel of a matrix (2.46) is spanned by a (diag b)-chain.
As an example for the application to integral operators we consider Sahnovich operators (4.9).

<u>COROLLARY 6.8.</u> The kernel of an operator (4.9) which is Fredholm [1], is spanned by a J-chain.

---

[1] This assumption can be replaced by dim ker A < ∞.

This statement follows from Theorem 6.3 and Corollary 4.1.
The last example will be singular integral operators.

COROLLARY 6.8. Let A be an operator of the form (5.3), which is Fredholm. Then ker A is spanned by $\alpha_0 \leq rn$ U_Γ-chains, where $U_\Gamma = tI$.

NOTES AND COMMENTS

Section 0.
For references see I. GOHBERG, N. KRUPNIK [2], S. MICHLIN, S. PRÖSSDORF [1].

Section 1.
1) The idea to exploit the reduction AU-VA for obtaining inversion formulae of A was used by L.A. SAHNOVICH [1-3] in the special case that U and V are integrations and by I. GOHBERG, G. HEINIG [3] in the case that U=V is the differentiation (c.f. also the paper of T. KAILATH and coauthors as well as A.B. NERSESJAN [1]). In its general abstract form the principle of UV-reduction was formulated first in G. HEINIG [8] (see also G. HEINIG, K. ROST [2,3]).
2) For references concerning the linear system theory see e.g. W.A. VOLOVICH [1], H. BART, I. GOHBERG, M.A. KAASHOEK [1].

Section 2 and 3.
1) Matrices close to Toeplitz were introduced and widely investigated in the papers of T. KAILATH, M. MORF and coauthors for the first time (see T. KAILATH, S. KUNG, M. MORF [1], B. FRIEDLANDER, T. KAILATH, M. MORF [1], B. FRIEDLANDER, T. KAILATH, L.LJUNG, M. MORF [1,2], S.LJUNG [1]). The approach of these authors differs from the approach presented here. The latter one is taken from G. HEINIG, K. ROST [2,3].
2) Matrices close to Vandermonde have been investigated in G. HEINIG, K. ROST [3].
3) Inversion algorithms for matrices with a small rank diagonal reduction are constructed for the first time.
4) The idea of transforming a Toeplitz plus Hankel matrix into a matrix of mosaic type is similar to an idea widely used in the theory of singular integral operators with Carleman shift (see e.g. D. PRZEWORSKA-ROLEWICZ [1]). Inversion formulae (different from our formulae) for Toeplitz plus Hankel matrices were constructed by A.B. NERSESJAN, A.A. PAPOJAN [1] (see also S. LJUNG [1]).
5) The inversion of paired T-matrices is considered more in detail in K. ROST [1].
6) The results and algorithms presented in these sections can be generalized to the block case.
7) It is an open problem how to construct the fundamental solutions for Toeplitz-like matrices if the matrix has non-regular principal sections.

Section 4.

1) The class of JJ-TLO is described for the first time.

2) Operators of the form (4.9) have been investigated by L.A. SAHNOVICH [1-3,5,6]. We note that in [5] there can be found further results about this class of operators and a lot of applications. Furthermore, our approach slightly differs from Sahnovich's one. The formulae and the proofs presented here are (in our opinion) simpler.

3) The class of DD-TLO coincides with the class of integral operators, which have been investigated by T. KAILATH, L. LJUNG, M. MORF [1]. Some generalizations of this class are considered by A.B. NERSESJAN [1].

4) Prop. 4.13 was proved with other methods in I. GOHBERG, A. SEMENCUL [1] and I. GOHBERG, G. HEINIG [3].

5) There exists a quite large number of papers concerning the continuous analogues of the recursion formulae of Section 3. These formulae are based on the well-known Bellman-Krein-Siegert identity for the resolvent kernel of an integral operator and are close to the method of invariant imbedding (see T. KAILATH, A. VIEIRA, M. MORF [1], I. GOHBERG, I. KOLTRACHT [1], J. CASTI, R. KALABA [1] and references there).

6) Another method for inverting of integral operators with displacement kernel was suggested by M. KREIN [2] (see also I. GOHBERG, M.KREIN [2]).

7) Operators with a generating function of the form (4.23) can be regarded as continuous analogues of Bezoutians. They have similar properties (see I. GOHBERG, G. HEINIG [5], L.A. SAHNOVICH [4]).

Section 5.

1) For references on singular integral operators see e.g. N.I. MUSKHELISHVILI [1], N.P. VEKUA [1], I. GOHBERG, N. KRUPNIK [2], S. MICHLIN, S. PRÖSSDORF [1], K. CLANCEY, I. GOHBERG [1]; for references on Toeplitz operators see I. GOHBERG, I. FELDMAN [1], F.D. GAHOV, J.I. CHERSKI [1], R.G. DOUGLAS [1].

2) The main result of this section, Theorem 5.2, was proved for Hölder continuous coefficients in G. HEINIG [6].

3) Theorem 5.2 can be generalized to the case of piecewise continuous coefficients.

4) Let us shortly compare the inversion formula (5.3) with the formula obtaining by the classical factorization method. First we note that the latter method gives a explicit formula in the case r=1 only. Suppose $A=dP+Q$ ($P+Q=I$, $P-Q=S_\Gamma$), and let $d=d_-d_+$ be a factorization of d. Then the solutions of the fundamental equations are

$$x = \frac{1}{2}(d_- - d_+^{-1}), \quad y = d_-^{-1}.$$

On the other side, the factorization factors are obtained via

$$d_- = y^{-1}, \quad d_+ = (y^{-1} - 2x)^{-1}.$$

Section 6.

1) The concept of partial indices is commonly used in the theory of

Toeplitz (Wiener-Hopf) and singular integral operators (see I. GOHBERG, M. KREIN [1], N.I. MUSKHELISHVILI [1], K. CLANCEY, I. GOHBERG [1] and references there). However, let us note that partial indices are usual associated with the symbol of the operator (i.e. with a matrix function) and not with the operator.

2) The concept of partial indices of an operator was introduced in G. HEINIG [10]. From this paper most of the results are taken. Theorem 6.8 is new. Let us note that Theorems 6.7 and 6.8 are even in the case of Toeplitz operators stronger than the classical results (c.f. I. GOHBERG, M. KREIN [1]).

3) The representation (6.17) can be used for further investigation of the kernel of TLO. Let us shortly explain this. We restrict ourselves to the case N=0. The relation (6.17) can be written in the form

$$(V - \Xi_1 G_1)A = A(U - \Gamma_2 X_2).$$

We introduce the transfer functions

$$\theta_1(\mu)=I_{r-\alpha}+\mu\,\Gamma_1(I-\mu U)^{-1}X_1,\quad \theta_2(\mu)=\mu\,\Gamma_2(I-\mu U)^{-1}X_1.$$

Then the kernel of A can be described by means of the common Jordan chains of θ_1 and θ_2. In particular, A is invertible if and only if $\sigma(\theta_1)\cap\sigma(\theta_2)=\emptyset$. In that way the Bezoutian concept can be generalized.

REFERENCES

Akaike, H.

[1] Block Toeplitz matrix inversion. SIAM J. Appl. Math. 24, 2 (1973), 140–156.

Akhiezer, N., and M. Krein

[1] On some questions of the theory of moments. (Russian) DNTVU, Charkov 1938.

Andersen, B., and E. Jury

[1] Generalized Bezoutian and Sylvester matrices in multivariable linear control. IEEE Trans. on a.c. AC-21, 4 (1976), 551-556.

Baxter, G., and I.I. Hirschman

[1] An explicit inversion formula for finite-section Wiener-Hopf operators. Bull. AMS 70 (1974), 820–823.

Barnett, S.

[1] A note on the Bezoutian matrix. SIAM J. Appl. Math. 22, 1 (1972), 84–86.

Barnett, S., and P. Lancaster

[1] Some properties of the Bezoutian for polynomial matrices. Lin. and Multilin. Alg. 9 (1980), 99–110.

Barnett, S., and C. Storey

[1] Some applications of the Lyapunov matrix equations. J. Inst. Math. Appl. 4, 3 (1968), 33–42.

Bart, H., I. Gohberg and M.A. Kaashoek

[1] Minimal factorization of matrix and operator functions. Birkhäuser Verlag, Basel-Boston-Stuttgart 1979.

Bitmead, R., and B. Anderson

[1] Asymptotically fast solution of Toeplitz and related systems of linear equations. Lin. Alg. and its Appl. 34 (1980), 103-116.

Böttcher, A., and B. Silbermann

[1] Invertibility and asymptotics of Toeplitz matrices. Akademie-Verlag, Berlin 1983.

Casti, J., and R. Kalaba

[1] Imbedding methods in applied mathematics. Addison-Wesley Publishing Company, Reading, Massachusetts, 1973.

Clancey, K., and I. Gohberg

[1] Factorization of matrix functions and singular integral operators. Birkhäuser Verlag, Basel-Boston-Stuttgart 1981.

Datta, B.N.

[1] Application of Hankel matrices to the root location problem. IEEE Trans. on a.c. AC-21, 4 (1976), 610–612.

Dickinson, B.W.

[1] Solution of linear equations with rational Toeplitz matrices. Math. of Computation 34, 149 (1980), 227-233.

Duduchava, R.

[1] Integral equations with fixed singularities. Teubner-Verlag, Leipzig 1979.

Durbin, J.

[1] The fitting of time-series models. Rev. Inst. Intern. Statist. 28 (1960), 233-244.

Douglas, R.G.

[1] Banach algebra techniques in operator theory. Acad. Press, New York 1972.

Fischer, E.

[1] Ueber das Caratheodory'sche Problem, Potenzreihen mit positivem reellem Teil betreffend. Rend. del Circ. Mat. di Palermo 31 (1911), 240-256.

Freund, G.

[1] Orthogonale Polynome. Deutscher Verlag der Wissenschaften, Berlin 1969.

Friedlander, B., T. Kailath and M. Morf

[1] A modified displacement rank and some applications. Proc. IEEE, Decision and Control Conf., New Orleans 1977, 958-961.

Friedlander, B., M. Morf, T. Kailath and L. Ljung

[1] New inversion formulas for matrices classified in terms of their distance from Toeplitz matrices. Lin. Alg. and Its Appl. 27 (1979), 31-60.

Frobenius, G.

[1] Ueber das Trägheitsgesetz der quadratischen Formen. Sitzungsber. d. Königl. Preuß. Akad. d. Wiss. (1894), 241-256, 407-431.

[2] Ableitung eines Satzes von Caratheodory aus einer Formel von Kronecker. Sitzungsber. d. Königl. Preuß. Akad. d. Wiss. (1912), 16-31.

Gahov, F.D., and J.I. Cherski

[1] Equations of convolution type. (Russian) Nauka, Moscow 1978.

Gantmacher, F. R.

[1] Matrizenrechnung I, II. Deutscher Verlag d. Wiss., Berlin 1970, 1971.

Gohberg, I., and I. Feldman

[1] Convolution equations and projection methods for their solutions. Math. Monogr. 41, Americ. Math. Soc., 1974.

Gohberg, I., and G. Heinig

[1] On the inversion of finite-section Toeplitz matrices. (Russian) Matem. Issled., Kishinev, 8, 3 (1973), 151-155.

[2] Inversion of finite-section Toeplitz matrices consisting of elements of a non-commutative algebra. (Russian) Rev. Roum. Math. Pures et Appl. 19, 5 (1974), 623-663.

[3] On matrix-valued integral operators on a finite interval with a kernel depending on the difference of the arguments. (Russian) Rev. Roum. Math. Pures et Appl. 20, 1 (1975), 55-73.

[4] The resultant matrix and its generalizations I. (Russian) Acta sci. math., Szeged, 37, 1-2 (1975), 41-61.

[5] Continuous analogue of the resultant operator II. (Russian) Acta Math. Acad. sci. Hung. 28, 3-4 (1976), 189-209.

Gohberg, I., and I. Koltracht

[1] Numerical solution of integral equations, fast algorithms and Krein-Sobolev equation. Preprint 1983.

Gohberg, I., and M. Krein

[1] Systems of integral equations on a half line with kernels depending on the difference of arguments. (Russian) Uspehi Mat. Nauk 13, 2 (1958), 3-72.

[2] Theory of Volterra operators in Hilbert space and its application. (Russian) Nauka, Moscow 1967.

Gohberg, I., and N. Krupnik

[1] A formula for the inversion of finite-section Toeplitz matrices. (Russian) Matem. Issled., Kishinev, 7, 2 (1972), 272-284.

[2] Einführung in die Theorie der eindimensionalen singulären Integraloperatoren. Birkhäuser Verlag, Basel-Boston-Stuttgart 1979.

Gohberg, I., and L.E. Lerer

[1] Resultant of matrix polynomials. Bull. AMS 82, 4 (1976), 565-567.

Gohberg, I., and A.A. Semencul

[1] On inversion of finite-section Toeplitz matrices and their continuous analogues. (Russian) Matem. Issled., Kishinev, 7, 2 (1972), 201-224.

Grenander, U., and G. Szegö

[1] Toeplitz forms and their applications. Univ. of California Press, Berkeley 1958.

Hartwig, R. E.

[1] Resultants and the solution of AX - XB = - C*. SIAM J. Appl. Math. 23 (1972), 104-117.

[2] AX - XB, resultants and generalized inverses. SIAM J. Appl. Math. 28, 1 (1973), 154-183.

Heinig, G.

[1] Endliche Toeplitzmatrizen und zweidimensionale Wiener-Hopf-Operatoren mit homogenem Symbol. I, II. Math. Nachr. 82 (1978), 29-68.

[2] On the concepts of Bezoutian and resultant for operator pencils. (Russian) Funkc. Anal. i Ego Pril. 11, 3 (1977), 94-95.

[3] Resultante, Bezoutiante und Spektralverteilungsprobleme für Operatorpolynome. Math. Nachr. 91 (1979), 23-43.

[4] Ueber Block-Hankelmatrizen und den Begriff der Resultante für Matrixpolynome. Wiss. Zeitschr. d. TH Karl-Marx-Stadt 19, 4 (1977), 513-519.

[5] Verallgemeinerte Resultantenbegriffe bei beliebigen Matrixbüscheln. I, II. Wiss. Zeitschr. d. TH Karl-Marx-Stadt 20, 6 (1978), 693-704.

[6] On the inversion of singular integral operators. (Russian) Bull. Acad. Sci. of Georg. SSR 96, 1 (1979), 29-32.

[7] Beiträge zur Spektraltheorie von Operatorbüscheln und zur algebraischen Theorie von Toeplitzmatrizen. Diss. B, TH Karl-Marx-Stadt 1979.

[8] Ein Prinzip zur Invertierung einiger Klassen linearer Operatoren. 7. TMP-Tagungsberichte II, TH Karl-Marx-Stadt 1979, 45-50.

[9] Transformation von Toeplitz- und Hankelmatrizen. Wiss. Zeitschr. d. TH Karl-Marx-Stadt 21, 7 (1979), 859-864.

[10] Struktur des Kernes und partielle Indizes bei allgemeinen linearen Operatoren. Preprint P-36/80, AdW der DDR, ZIMM, Berlin 1980.

[11] Inversion of Toeplitz and Hankel matrices with singular sections. Wiss. Zeitschr. d. TH Karl-Marx-Stadt 25, 3 (1983), 326-333.

Heinig, G., and U. Jungnickel

[1] Zur Lösung von Matrixgleichungen der Form AX - XB = C. Wiss. Zeitschr. d. TH Karl-Marx-Stadt 23, 4 (1981), 387-393.

[2] The Routh-Hurwitz problem and the Bezoutian for matrix polynomials. Math. Nachr. (to appear).

Heinig, G., and K. Rost

[1] Ueber homogene Gleichungen vom Faltungstyp auf einem endlichen Intervall. Demonstratio mathem. 10, 3-4 (1977), 791-806.

[2] Invertierung einiger Klassen von Matrizen und Operatoren I. Endliche Toeplitzmatrizen und ihre Verallgemeinerungen. Wiss. Informat. 12, Karl-Marx-Stadt 1979.

[3] Invertierung von Toeplitzmatrizen und ihren Verallgemeinerungen, I. Die Methode der UV-Reduktion. Beiträge zur Num. Math. 12 (1983), 55-73.
II. Rekursive Algorithmen (unpublished).

Heinig, G., and B. Silbermann

[1] Factorization of matrix functions in algebras of bounded functions. Preprint P-Math-05/83, AdW d. DDR, IMath., Berlin 1983.

Hirschmann Jr., I.I.

[1] Finite section Wiener-Hopf equations on a compact group with orderd dual. Bull. AMS 70 (1964), 508-510.

[2] Matrix-valued Toeplitz operators. Duke Math. J. 34, 3 (1967), 403-416.

Householder, A.S.

[1] Bezoutians, elimination and localization. SIAM Review 12, 1 (1970), 73-78.

Iohvidov, I.S.

[1] On the signature of Toeplitz forms. (Russian) DAN SSSR 169, 6 (1966), 1258-1261.

[2] On Hankel and Toeplitz matrices and signatures of Toeplitz forms. (Russian) Ukr. Matem. Žurn. 19, 1 (1967), 25-35.

[3] On the rank of Toeplitz matrices. (Russian) Matem. Sborn. 76, 1 (1968), 26-38.

[4] On the rules of Jacobi, Gundelfinger and Frobenius. Matem. Issled., Kishinev, 3, 4 (1968), 162-165.

[5] On the (r,k)-characteristic of a Hankel matrix. (Russian) Uspehi Mat. Nauk 26, 4 (1969), 199-200.

[6] On Hankel matrices and forms. (Russian) Matem. Sborn. 8, 2 (1969), 241-252.

[7] On the Fischer-Frobenius transformation. (Russian) Teorija Funkc., Funkc. Anal. i Ih Pril., Charkov, 15 (1972), 203-212.

[8] Hankel and Toeplitz matrices and forms. (Russian) Nauka, Moscow 1974.

[9] On the (r,k)-characteristics of rectangular Hankel matrices. (Russian) DAN SSSR 233, 2 (1977), 276-279.

Iohvidov, I.S., and M. Krein

[1] Spectral theory of operators in spaces with an indefinite metric II. Trudy Mosk. Mat. Obsch. 8 (1959), 413-496.

Iohvidov, I.S., and O.D. Tolstyh

[1] On the (r,k,l)-characteristics of rectangular Toeplitz matrices. Ukr. Matem. Ž. 32, 4 (1980), 477-482.

Kailath, T., S.-Y. Kung and M. Morf

[1] Displacement ranks of matrices and linear equations. J. Math. Anal. Appl. 68, 2 (1979), 395-407.

Kailath, T., B. Levy, L. Ljung and M. Morf

[1] The factorization and representation of operators in the algebra generated by Toeplitz operators. SIAM J. Appl. Math. 37, 3 (1979), 467-484.

Kailath, T., L. Ljung and M. Morf

[1] Generalized Krein-Levinson equations for efficient calculation of Fredholm resolvents of non-displacement kernels. Topics in Funct. Anal., Acad. Press, 3 (1978), 169-184.

Kailath, I., A. Vieira and M. Morf

[1] Inverses of Toeplitz operators, innovations, and orthogonal polynomials. SIAM Rev. 20 (1978), 106-119.

Koltracht, I.

[1] Inversion formulas for operators which are close to Toeplitz operators. Int. Equ. and Oper. Theory 5, 1 (1982), 71-84.

Kravitsky, N.

[1] On the discriminant function of two commuting non selfadjoint operators. Operator Theory and Int. Equ. 3 (1980), 97-124.

Krein, M.

[1] Integral equations on the half-line with a kernel depending from the difference of arguments. (Russian) Uspehi Mat. Nauk 13, 5 (1958), 3-120.

[2] On a new method for solving linear integral equations of the first and second kind. (Russian) DAN SSSR 100, 3 (1955), 413-416.

Krein, M., and M. Neimark

[1] The method of symmetric and Hermitian forms in the theory of root separation of algebraic equations. (Russian) DNTVU, Charkov 1936.

Kutikov, L.M.

[1] Inversion of correlation matrices and some related questions. (Russian) Izv. Vuzov, Ser. Tehn. Kibernetika 5 (1965), 42-47.

[2] On the structure of matrix inverses to correlation matrices of vector-valued random processes. (Russian) Ž. Vyč. Mat. i Mat. Fiz. 7, 4 (1967), 764-773.

Lander, F. I.

[1] The Bezoutian and the inversion of Hankel and Toeplitz matrices. Matem. Issled., Kishinev, 9, 2 (1974), 69-87.

Lerer, L., and M. Tismenatsky

[1] The Bezoutian and the eigenvalue separation problem for matrix polynomials. Int. Equ. and Operator Theory 5, 3 (1982), 386-445.

Levinson, N. L.

[1] The Wiener RMS (root mean square) error criterion in filter design and prediction. J. Math. and Phys. 25 (1947), 261-278.

Ljung, S.

[1] Fast algorithms for integral equations and least square identification problem. Linköp. Stud. in Scien. and Techn., Diss. 93, 1983.

Müller, P. C.

[1] Solution of the matrix equations AX-BX = -Q and S^TX-XS = -Q. SIAM J. Appl. Math. 18 (1970), 104-117.

Muskhelishvili, N. I.

[1] Singular integral equations. (Russian) Nauka, Moscow 1968.

Michlin, S., and S. Prößdorf

[1] Singuläre Integraloperatoren. Akademie-Verlag, Berlin 1980.

Nersesjan, A. B.

[1] The structure of resolvent of some integral operators. (Russian) Izv. AN Arm. SSR 17, 6 (1982), 442-436.

Nersesjan, A.B., and A.A. Papojan

[1] Construction of matrix inversed to the sum of Toeplitz and Hankel matrices. (Russian) Izv. AN Arm. SSR 18, 2(1983), 150-160.

Nussbaumer, H. J.

[1] Fast Fourier transform and convolution algorithms. Springer Verlag, Berlin-Heidelberg-New York 1981.

Postnikov, M. M.

[1] Stable polynomials. (Russian) Nauka, Moscow 1981.

Pozin, S. M., and L. A. Sahnovich

[1] On the solution behaviour of the equation with displacement kernel. Izv. AN Arm. SSR 17, 5 (1982), 375-386.

Prößdorf, S.

[1] Einige Klassen singulärer Gleichungen. Akademie-Verlag, Berlin 1974.

Przeworska-Rolewicz, D.

[1] Equations with transformed arguments, an algebraic approach. PWN, Warszawa 1973.

Ptak, V., and N. J. Young

[1] A generalization of the zero location theorem of Schur and Cohn. Lin. Alg. and Its Appl. (to appear).

[2] Zero location by Hermitian forms: the singular case. Lin. Alg. and Its Appl. (to appear).

Rissanen, J.

[1] Algorithms for triangular decomposition of block Hankel and Toeplitz matrices with application to factoring positiv matrix polynomials. Math. of Comput. 27, 121 (1973), 147-154.

Rost, K.

[1] Invertierung einiger Klassen von Matrizen im Zusammenhang mit der näherungsweisen Lösung von Operatorgleichungen. Diss. A, TH Karl-Marx-Stadt 1980.

Russakovski, E. M.

[1] On Hankel and Toeplitz matrices and the Bezoutian I. (Russian) Teorija Funkc., Funkc. Anal. i Ih Pril., Charkov, 32 (1979), 77-82.

[2] —— . II. 33 (1980), 119-124.

Sahnovich, A. L.

[1] On a method for inverting finite-section Toeplitz matrices. (Russian) Matem. Issled., Kishinev, 8, 4 (1973), 180-186.

Sahnovich, L. A.

[1] On similarity of operators. (Russian) Sibirskij Mat. Ž. 8, 4 (1972), 868-883.

[2] On the integral equation with a kernel depending on the difference of arguments. (Russian) Matem. Issled., Kishinev, 8, 2 (1973), 138-146.

[3] On a class of integral equations solvable in closed form. (Russian) Matem. Issled., Kishinev, 9, 1 (1974), 157-164.

[4] The Bezoutian operator in the theory of root separation of entire functions. (Russian) Funkc. Anal. i Ego Pril. 10, 1 (1976), 62-66.

[5] Equations with displacement kernel on a finite interval. (Russian) Uspehi Mat. Nauk 35, 4 (1980), 69-129.

[6] Systems of equations with displacement kernel. (Russian) Ukr. Mat. Ž. 32, 1 (1980), 61-68.

Semencul, A. A.

[1] Inversion of finite sections of paired and transposed paired operators. (Russian) Matem. Issled., Kishinev, 3, 1 (1968), 100-107.

Smith, R. A.

[1] Matrix calculations for Lyapunov quadratic forms. J. Diff. Equ. 2 (1966), 208-217.

Trench, W. F.

[1] An algorithm for the inversion of finite Toeplitz matrices. SIAM J. Appl. Math. 12, 3 (1964), 515-522.

Vekua, N. P.

[1] Systems of singular integral equations. (Russian) Nauka, Moscow 1970.

Volovich, W. A.

[1] Linear multivariate systems. Springer Verlag, New York-Heidelberg-Berlin 1974.

Whittle, P.

[1] On the fitting of multivariate autoregressions and the approximate factorization of a spectral density matrix. Biometrika 50 (1963), 129-134.

Zohar, S.

[1] Toeplitz matrix inversion: the algorithm of W.F. French. J. Assoc. Comput. Math. 16 (1967), 592-601.

SUBJECT INDEX

NOTATION INDEX

Band 1

Inverse and Improperly Posed Problems in Differential Equations

Proceedings of the Conference on Mathematical and Numerical Methods, held in Halle/Saale (GDR) from May 29 to June 2, 1979

Edited by Gottfried Anger

296 Seiten, 1979

Band 2

Fundamentals of Computation Theory FCT '79

Proceedings of the Conference on Algebraic, Arithmetic and Categorial Methods in Computation Theory, held in Berlin/Wendisch-Rietz (GDR) September 17—21, 1979

Edited by Lothar Budach

576 Seiten, 1979

Band 3

Decidability and Generalized Quantifiers

By Andreas Baudisch, Detlef Seese, Hans-Peter Tuschik, and Martin Weese

XII, 235 Seiten, 1980

Band 4

Zur Elastizitäts- und Thermoelastizitätstheorie

Potentialtheoretische Methoden in Anwendung auf verschiedene Modelle

Von Lothar Jentsch und Johannes Maul

157 Seiten, 1980

Band 5

Systems Analysis and Simulation 1980

Proceedings of the International Symposium held at Berlin (GDR) September 1—5, 1980

Edited by Achim Sydow

496 Seiten, 1980

Band 6

Nichtlineare diskrete Optimierung

Von Eberhard Girlich und M. M. Kowaljow

218 Seiten, 1981

Band 7

Completeness Problems in the Structural Theory of Automata

By Jürgen Dassow

148 Seiten, 1981

Band 8

Darstellung von Lösungen linearer elliptischer Differentialgleichungen

Von Günther Wildenhain

92 Seiten, 1981

Band 9

Introduction to the Spectral Theory of Operators in Spaces with an Indefinite Metric

By I. S. Iohvidov, M. G. Krein, and H. Langer

120 Seiten, 1982

Band 10

Nonlinear Time-discrete Systems

A General Approach by Nonlinear Superposition

By Michael Gössel

115 Seiten, 1982

Band 11

Preprimal Algebras

By Klaus Denecke

162 Seiten, 1982

Band 12

Ergodic Theory and Related Topics

Proceedings of the Conference held in Vitte/Hiddensee (GDR) October 19—23, 1981

Edited by Horst Michel

232 Seiten, 1982

Band 13

Singularly Perturbed Differential Equations

By H. Goering, A. Felgenhauer, G. Lube, H.-G. Roos, and L. Tobiska

176 Seiten, 1983

Band 14

Leben wir in einer Volterra-Welt?

Ein ökologischer Zugang zur angewandten Systemanalyse

Von Manfed Peschel und Werner Mende

233 Seiten, 1983

Band 15

Markov Decision Problems with Countable State Spaces Optimality Criteria — Algorithms — Clustering

By Hans Michael Dietz and Volker Nollau

174 Seiten, 1983

Band 16

Stochastische Verfahren zur Suche nach dem Minimum einer Funktion

Von Ryszard Zieliński und Peter Neumann

133 Seiten, 1983

Band 17

Invertibility and Asymptotics of Toeplitz Matrices

By Albrecht Böttcher and Bernd Silbermann

200 Seiten, 1983

The first part deals with matrices having a Toeplitz structure $[a_{i-k}]$ or a Hankel structure $[b_{i+k}]$. An algebraic theory of these matrices is presented, which allows the construction of effective algorithms for the inversion and generalized inversion of these matrices. Furthermore, the theory gives an insight into the structure of Toeplitz and Hankel matrices and may be applied to root localization problems as well as to the solution of special matrix equations. It covers classical results on Toeplitz and Hankel matrices and presents new results published here for the first time. In the second part it is shown that the approach offered in the first part can be applied to further classes of matrices and linear operators, in particular to matrices close to Toeplitz or Vandermonde, to Wiener-Hopf operators, to singular integral operators and to some Fredholm integral operators.
The volume is written for mathematicians interested in linear algebra, especially numerical problems, for specialists in the fields of operator theory and integral equations as well as for specialists in system and control theory.

Der erste Teil ist Matrizen gewidmet, die eine Toeplitzstruktur $[a_{i-k}]$ oder Hankelstruktur $[b_{i+k}]$ besitzen. Es wird eine algebraische Theorie entwickelt, mit deren Hilfe effektive Algorithmen zur Invertierung und verallgemeinerten Invertierung dieser Matrizen konstruiert werden können. Darüber hinaus gibt die Theorie Einblick in die Struktur von Toeplitz- und Hankelmatrizen und besitzt Anwendungen bei Nullstellenverteilungsproblemen algebraischer Polynome sowie bei der Lösung von speziellen Matrixgleichungen. Sie liefert neben klassischen Resultaten über Toeplitz- und Hankelmatrizen neue Ergebnisse, die hier erstmalig veröffentlicht werden.
Im zweiten Teil wird gezeigt, daß sich der im ersten Teil dargelegte Zugang auf weitere Klassen von Matrizen und linearen Operatoren anwenden läßt, z. B. Toeplitzartige und Vandermondeartige Matrizen, Wiener-Hopfsche und singuläre Integraloperatoren sowie gewisse Klassen Fredholmscher Integraloperatoren.
Der Band ist geschrieben für Mathematiker, die an (u. a. numerischen) Problemen der linearen Algebra interessiert sind, für Spezialisten auf den Gebieten Operatorentheorie und Integralgleichungen sowie der linearen System- und Kontrolltheorie.

ISSN 0138-3019

MIX
Papier | Fördert gute Waldnutzung
FSC® C083411

Zeitfracht Medien GmbH
Ferdinand-Jühlke-Straße 7
99095 Erfurt, Deutschland
produktsicherheit@kolibri360.de